THE TUNGUS EVENT

THE TUNGUS EVENT
OR THE GREAT SIBERIAN METEORITE

John Engledew

Algora Publishing
New York

Library of Congress Cataloging-in-Publication Data —

Engledew, John, 1954-
    The Tungus event, or, the great Siberian meteorite / John Engledew.
       p. cm.
    Includes bibliographical references and index.
      ISBN 978-0-87586-781-6 (case laminate: alk. paper) — ISBN 978-0-87586-780-9
(trade paper: alk. paper) — ISBN 978-0-87586-782-3 (ebook) 1. Tunguska meteorite. I.
Title. II. Title: Tungus event. III. Title: Great Siberian meteorite.
    QB756.T8E54 2010
    551.3'9709575—dc22
                          2010000102

Dedicated to Tamara, Devin and all my wonderful friends.

TABLE OF CONTENTS

PREFACE AND A COSMIC VIEW

The profoundly mysterious object from space that exploded several miles above a forested expanse of Siberia early one summer morning in 1908 defies full explanation in the face of a century of investigation.

That it was a major meteor or piece of a comet entering the Earth's atmosphere and powerfully erupting above ground level is still the most common description, but theories exist in profusion. They include rogue asteroids, mini black holes and even alien intervention. As we will see, these explanations are not entirely equal to the facts. In the unique case of the Tunguska event, there was wholesale destruction to the mighty taiga woodlands but none of the debris that one would expect should exist from the body itself. Evidence like a strewn field of meteorite debris or meteoric dust on the trees and ground were never found, nor were any craters, in the area beneath the site of the fireball nor anywhere along the path it took. There are no craters because the Tunguska Cosmic Body (TCB) did not hit the ground. Atmospheric anomalies prior to the dramatic appearance of the fiery body puzzlingly occurred for several days, adding to the enigma.

After reading several English language books and numerous articles on the subject, I decided to pull together an up to date and succinct study of the strange and unexplained events with some background in meteor, comet and astronomical studies to try and put it all in perspective. This project, originally intended to be a set of notes culled from various sources and composed for personal use, soon became a labor of love. Separately, the protagonists of pseudoscience may be genuine in effort and everyone is entitled to an opinion, but for our purposes I aim to present just what is accepted as rigorously scientific. I've waded through plenty of material that is quite fabulistic, and that includes energetic blogs concerning Tunguska. The curse of insufficient data has long haunted Tunguska studies.

I've been caught up in the wonder of the universe for years and hope to convey something of the grandeur of cosmic nature to others. It has dynamism and beauty that words cannot describe. The progress we have made in recent decades in comprehending the universe is a phenomenon in itself. Amateur astronomy is a rewarding pursuit with the simplest optical equipment or just eyesight revealing great wonders in the night sky. The Tungus event is one of the astronomical mysteries that is tangible and immediate because it happened on the Earth. We still lack a full understanding of what occurred that day, but we can examine it in the context of astronomy and Earth science, local eye-witness accounts, and the findings of the early and subsequent scientific expeditions.

Like any investigation true to the spirit of science, I thought that reviewing the existing studies must reveal a solution to the puzzle. But the mysteries of the Tungus event actually multiply, the more one looks into it. One learns about ballistic waves, Soviet academic infighting, Near Earth Objects, native Evenki beliefs, cometary orbits and much else on the trail of Tunguska.

In the history of astronomy and astrophysics we have made some sturdy achievements, and some notable people have emerged who greatly added to the repository of human knowledge. We have observed, calculated and wrested powerful information on what were considered mystical questions in our cultural past. Such basic questions as the age of the Earth, the distance of the Moon and the composition of the Sun have been answered: 4.5 billion years, a quarter of a million miles, and hydrogen and helium in nuclear flux.

In the case of the great Siberian meteorite (no one has proposed any better name for it for decades) I fear that we are no closer to an unequivocal solution even today. One can have moments of clarity when it appears the case might have been cracked, at least to one's own satisfaction. All too soon, objections arise, facts contradict and one finds one's intellect pondering other possibilities. There are diverse and very different explanations. One is tempted to accept that it was an unusual meteor or piece of a comet and forget about aspects that stubbornly refuse to fit. One reads papers and articles with expressing various levels of confidence or brazenly claiming to have revealed all. I prefer the purist scientific ones and it takes a major foray into popular science writing to interpret the evidence meaningfully for the general reader.

In a tight summary of the consensus of explanations: There are asteroid or comet theories, the spaceship hypothesis, or a gaping unknown. The question is still wide open. Many people have heard of the great Siberian meteorite, but their main impression remains that "They never really figured out what it was, did they?" Most references describe the imposing scale of the damage and attribute matters to a giant meteor or small comet, adding that a mystery remains.

The original instigator of scientific study in the 1920s, the Estonian-born Leonid Alekseyevich Kulik (1883–1942) is a kind of Galileo of the subject. He produced a terrific analysis based on the limited resources in the 1920s and must be seen as the original pioneer and founding father of the subject. We owe much to his intellectual tenacity and physical endurance for exploring the site and circumstances of the event. Four decades later, the Volga German-Russian Wilhelm Fast (1936–2003) has been called its Newton. This is justly deserved considering his robust mathematical approaches and avoidance of hypothesizing. He entirely left that to others in the field of research.

We await the Einstein of Tunguska studies and I make no pretense of fulfilling that role. There are sensationalist websites that occasionally make absurd claims and connections, including the "spaceship theory." It is hard for humans to tolerate the existence of something we simply cannot account for, and for many of us the hunt for working hypotheses is irresistible. I hope in this book to have emerged with something worthwhile that will be an enjoyable, intelligent, stimulating read.

Our understanding of the solar system has been revolutionized in the last half century. The explosion of knowledge unfortunately does not include that oddly huge one in Siberia a century ago.

Geniuses from practicing scientists like Albert Einstein to those whose writings combine a scientific grounding with imaginative projection, like Carl Sagan, Isaac Asimov and Arthur C. Clarke, weigh in with a 2–1 vote for the comet theory followed by a specific meteor stream as causing the Tungus event.

I find that Clarke is unequalled in his imagination and sheer knowledge of cosmic nature. It is a potent combination. He postulates that the very date of June 30, the climax of the Beta Taurid meteor stream, quite simply indicates the physical nature of the Siberian fireball prior to falling to Earth. It must have been a peculiar Beta Taurid meteor and, if so, there is a wealth of assumed information on its prior paths in planetary space. In turn, the Comet Encke had spawned other closely related meteor streams, so Tungus was, likely, a piece of the Comet Encke. Does that give us an open and shut case just based on the date of the occurrence? There are many other meteor streams with predictable dates for their peaks. The Leonids and Perseids are worth waiting for beneath the starry expanse on specific dates and are quite reliable as annual spectacles.

I chose this trio as my mentors. Clarke's unbridled literary power inspired me from an early age. I can remember staying awake late into the night, engrossed in some piece of his. I also owe something to Isaac Asimov who, as one of the most prolific popular science writers of all time, was a truly great teacher. He conferred a certain "of course, you can understand this" sense upon diverse branches of science. Blending fact with some clever science fiction, he was as great an entertainer as educator. Then, the Sagan-Schlovskii collaboration, *Intelligent Life in the Universe*, will remain the definitive book on the subject for some time. Carl Sagan's "Cosmos" TV series and his explorations on the evolutionary history of the human being as a species are always worth revisiting.

If a "starship" is a vehicle that journeys to the stars, we have deliberately launched no fewer than four in the form of space probes released from the gravitational grasp of both the Earth and the Sun. The two Pioneer and two Voyager spacecraft are moving too fast for the Sun's great field of attraction to hold them and will exit the solar system out into the interstellar night, never to return. It would be marvelous if someone came across them and perused our careful communicative plaques and the information packages deliberately conveyed with the spacecraft. By radio noise, man has by now loudly announced his presence over a radius of 80 light years. A technological civilization would merely need equipment at our level of development to pick up the signals and watch them oscillate from one side to another around a certain visible star in a mere six months.

There are many major mysteries in nature. They still include the physiological function of the human body and brain, where our knowledge is anything but complete. The scholarly compartments of chemistry, physics or biology as separate disciplines are purely arbitrary. I prefer to focus on astronomical questions and approach them strictly from a scientific angle. The scales of cosmic time and space have long fascinated me and provide, at least on a personal level, some sense of perspective for our existence.

In recent years we have forged powerfully ahead. There is, for specific example, progress at last in understanding the origin of the dwarf planet Pluto being one of a class of objects at the extreme orbital distances from the Sun. There is indeed at least one body larger than Pluto lying beyond it, as was long speculated. A thousand times beyond the Kuiper Belt lies the Oort Cloud. It comprises the raw material of untold millions of comets lying in potential existence if they were ever beckoned sunward by a gravitational nudge.

There may be evidence that elementary life once existed on Mars but did not endure due to the changing and inhospitable environment that took over there. Enormous importance could be placed on such a possibility. Whatever policies and budgeting are decided for future space research, some unmanned missions travelling in space are ongoing, as we speak, with informative results to look forward to. This includes scheduled comet encounters and a flyby of the far flung Pluto, the only major member of the solar system yet to be imaged close up, coming hopefully in July 2015. An extended mission to pay premier visits to Kuiper Belt Objects is slated for this same mission.

Whether the United States will be making a return to the Moon or launching a manned mission to Mars anytime in the next decades remains to be seen. Both look increasingly unlikely. In fact the United States, on the retirement of its Space Shuttle, does not have a proven and assembled booster to do any job of that sort. It must be said that what NASA pulled off in the first twenty years since its inception in the late 1950s cannot have been easy technically or, I suppose, politically. Apollo's nine manned missions to the Moon were mountaintop experiences as were the two flights in Earth orbit. We set foot there six times, making an exclusive Moon walking club of twelve American astronauts. In the spirit of Tsiolkovsky, Goddard, and Von Braun whose rocket-propelled dreams all came true, it has been said that the rocket changed the future of the world, indeed of many worlds. Those pioneers literally got us off the ground.

Konstantin Tsiolkovsky (1857–1935), a Russian schoolteacher, made some astute mathematical drawing-board speculations on the mechanics of putting hardware into orbit. He worked in comparative obscurity before the Wright brothers constructed their flying machines and first took to the air by powered flight in 1903.

The German engineer Otto Lilienthal had successfully built and flown piloted gliders and in 1889 published a book on aerodynamics, a technical work consulted by the Wright brothers. Others in Europe were also experimenting with various types of flying machines. Samuel Langley achieved a working steam powered model plane that flew ¾ mile in 1891 but was unable to construct a full-scale aircraft using similar principles and propulsion. In the United States during the 1930s Robert Goddard built small but successful chemically-fueled rocket models and Von Braun's Saturn V booster ultimately reached the Moon. Concerning astronautics, the Apollo 8 astronauts burned the ship's rocket motors to higher than 25,000 miles an hour for trans-lunar injection in December 1968. This was the first time a human crew ever struck out from Earth's orbit toward another celestial body, and it was a momentous event — not merely in manned spaceflight but in human history.

Deciding exactly how one spends the tiny and temporary flicker of rational consciousness with which we are endowed is some sort of existential human right, I assume. Recognizing that there is such an entity is not universal. There is much to Descartes' "cogito ergo sum" ("I think, therefore I am") and the Platonic advice that the unexamined life is not worth living. Sometimes merely standing beneath the starry sky and reading skypub.com like a news service reviewing the cutting edge of astronomy must suffice. The sweep of the seasonally shifting constellations, Moon and planets is one of the most glorious and inexhaustible sights available to us all. It is woefully underrated as a spectacle. You have the universe in your backyard.

The Tungus event is a tangible, "right here" kind of mystery with the door forever open to new studies and reinterpretation. Many major meteor falls have left the tangible evidence of physical pieces and craters. Their presence on the Earth's surface is sometimes so large it took centuries for humans to recognize them. Tunguska is truly singular; nothing like it appears to have happened before or since.

The event could have been soon forgotten, reduced to some indigenous legend in a scary Siberian folktale. It immediately took on a fearsome mystique as early researchers found when they met a wall of silence on the matter from the local Evenki people.

By applied research, humans have cracked tougher scientific problems. Serendipitously, the accidental discovery of cosmic microwave radiation was a big step in understanding the origin of the universe and an example of what can be stumbled upon. So far, this kind of technical key remains to be found that can explain Tungus satisfactorily.

Two aspects of the event capture the imagination. Like a giant Rubic's cube, it has thus far defied resolution. It seems an insuperable challenge to juggle all the observations and interpretations concerning the Tungus event and emerge with a viable solution that accords with all the facts and is acceptable to all factions. Second, whatever the scientific explanation of what happened that day, it is never recognized that the human race escaped a major disaster due only to the sheer remoteness of the region where it all took place. Unlike the scientific and environmental vacation trips that promise to take tourists to the right place for viewing total eclipses, there is no package deal promising a "journey to Tunguska." Until we see that advertised, we have to make do with a very good You Tube mini documentary: the epicenter and Lake Cheko are filmed from a Russian army helicopter. The riddle is important to resolve, but it continues to taunt and humble those who take satisfaction in mankind's ability to crack nature's riddles.

Eye-Witness Accounts

Let us begin with the evaluated testimonial evidence, namely the eye-witness accounts from the Tungus people and Russian farmers, herdsmen and hunters that day, as read and analyzed almost into dust. A remote part of central eastern Siberia was the scene, early in the morning. Here are the broad facts of the case.

Tunguska is a general geographic name for a part of central eastern Siberia in post-Soviet Russia, taking its name from rivers in the region. The term has since attached itself to the greater mystery.

A body from space arrived, in the general sense small and very bright, round or possibly cylindrical in shape. It may have appeared both larger and brighter than the Sun as it blazed from the SE to the NW across the sky for 5–10 minutes before exploding. The bright tail was iridescent, conveying a wide range of colors as reports come down to us. The fact that it was observable at all, descending from the sky for that time, indicates a comparatively low speed. This can be deduced from the simple rates of motion for the relatively low key and more familiar cosmic events like meteors that are going on all the time. There were powerful explosions and crashes, none of which would have been a sonic boom — it is not thought that the Tunguska Cosmic Body decelerated to beneath the speed of sound. There is not, even now, full agreement that it slowed down at all.

The path and speed of the body cannot be confidently calculated, but investigators tend to assume the lower limits for velocity. We cannot assess the order or sizes or decisively pin down the number of bangs. Much of the data is infuriatingly vague. I have not included any detailed work done on trying to map its course. The best we can do is to speculate that it sped from somewhere north of

Lake Baikal in eastern central Siberia over a trajectory of some 400 ground miles to a densely forested spot 37 miles north of a small settlement called Vanavara.

Let us firstly compare the projectile's speed with the velocities of those ordinary meteors regularly burning up in the atmosphere. Here we have plenty of examples to debate. This is a very well observed daily occurrence. The pace was likely on the order of half a mile a second or slightly more than some 40 miles a second as measured for faster incoming meteors. This is simply concluded because there was time to see it coming. It may have come out of the Sun, from the direction of the Sun, or from a little to the west of it. Which of those is accurate could alone prove to be a crucial fact in helping to determine what it was and how it happened. And had it packed a higher cosmic velocity, there would have been no warning or drawn-out sighting at all.

Or perhaps, it was not seen coming from out in space earlier in the clear dawn because it initially travelled so fast. Equally possible, it was a compact, dark projectile unseen in the glare of the Sun. It proceeded to slow down enormously in its passage downward through the Earth's atmosphere, as meteors may do, and only glowed hot on the very final passage. These are real but very different possibilities for its approach.

There is always a risk that a small, dark, high speed meteor could hit Earth without notice. We need to understand potential threats to the planet, the biosphere and civilization. The Jet Propulsion Laboratory's Near Earth Asteroid Tracking (NEAT) project is one important effort to identify and keep a tab on any such anomalies. Their web pages show how much has been accomplished in observing Near Earth Objects. In early 2010 a new report from the National Research Council describes the options for NASA to detect more NEOs. In 2005 Congress had mandated that NASA discover 90% of such bodies whose diameter is 140 meters or greater by 2020.

A low angle of approach and comparatively slow speed are generally accepted for the mysterious space body explored in this book. It was seen over an area equivalent to the land areas of France and Germany combined. This indicates more of a horizontal than vertical trajectory, and it was accompanied by some frightful noises as it flew low enough for acoustic waves to be created. One of the many interpretations among some startled onlookers was that the end of the world was at hand.

There are some unknowns in the propagation of sound associated with meteor paths but here it was loud and clear. The curvature of the Earth and the horizon restricting the view of a low-flying object further afield further supports the case for a low angle of approach to the ground, at least on the final leg of its journey. This is all we can assess by the incidental observations with a few incisive points noted here and there by individual witnesses. I will quote such accounts to give the reader a feel for the strange apparition of 30 June 1908. The precise path from its first sighting cannot be properly worked out. There are, lamentably, contradictions thrown up by their close analysis. The TCB may have altered course more than once and, as a remote possibility, there could have been more than one body in flight.

A certain amount of information can be calculated, using broad estimates of the speed and angle of approach over a few hundred terrestrial miles and the brief window of visual contact for this object exploding out of the sky. The math is simple and I am not a patronizing expert in the subject. The damage to the forest below the explosion would have been more elliptical or linear in shape had it borne a higher speed. Analysts unequivocally agree that the TCB somehow engulfed itself in a powerful fireball at a height of about five miles. There lies the profoundest of mysteries. What sort of chemical or nuclear explosion took place?

Intense heat and powerfully hot air waves were unleashed at the final fall point. The sheer power of the blast remains one of the major unsolved problems, and the strange and huge damage to the woodlands was quite unlike that of a normal forest fire or a hurricane's path. A large radial pattern flattened trees like so many matchsticks around a central zone, over 800 square miles in extent. The blast and heat were intense yet brief and quite unlike a normal woodland conflagration. The trees were charred and singed, generally far more on the side facing the central part of the fall site. They were not burned like they would be in any ordinary forest inferno (which are, incidentally, relatively uncommon in the snowy Siberian taiga but can occur in dry summers).

What could have caused it? A meteor would have left fragments or some kind of debris field, at least meteoric dust, strewn beneath its path or at the final place of detonation. If the cosmic powder was that fine then microscopes eventually would have revealed them in soil and peat samples somewhere from the area. Many tests have been done to seek such evidence, but they remain inconclusive.

A meteor probably could not have supplied such huge explosive force in the first place. A large space rock cannot pack such mega-tonnage of explosion nor the order of heat necessary to disperse itself completely when merely warmed a bit more by the friction of atmospheric entry. Such statements are, at least, primary points of technical contention in the physics of ballistics.

There must have been an inner source of energy to the body, and here is the crux of the matter. This is what cannot be fathomed. Following the explosion, we have the giant riddles of traceless Tunguska on the ground apart from the destruction and heat charring of the forest that was rendered. Exceptional, great white nights followed as Russia and Western Europe beheld long, bright twilights for weeks afterwards. The clarity of the atmosphere was diminished significantly for weeks. A comet or cometary fragment would have burned up and dissipated far higher in the atmosphere than the altitude where the explosion occurred. So then, what was it?

One practical reason for trying to understand the mystery would be to anticipate and ideally counter something like it ever happening again. Some sort of major impact from a meteor is bound to take place in the future. Asteroids are simply bigger versions and there is no significant difference apart from size. We will see how important they have proven in the Earth's past, with biological fallout, as it were, as major impacts have interspersed the progress of life on Earth — destructively, in the short run. The death of the dinosaurs was a scenario whose reality was brilliantly unearthed and expounded from 65 million years ago by Luis Alvarez and team. There is full supporting geological evidence that their extinction was the result of a major asteroid impact. We've even located the buried crater. Mass extinctions have occurred several other times in the greater past.

The paths of more immediate human history, at least in Europe or Russia, would have been tragically altered in one instant had the object erupted over a major center of population there. The knowledge for which we pride ourselves concerning Near Earth Objects still cannot prevent a smaller "big" meteor appearing with minimal warning. A familiar science fiction plot revolves around the alarming discovery of a major asteroid heading for Earth. How do we react and deal with it?

We've seen movies starring heroic astronauts bravely disrupting the meteor's course or blowing it to pieces out in space, narrowly preventing a devastating series of blows to Earth from giant impacters. Other tales describe the post-impact world of nuclear winter and a serious dislocation of civilization and human existence itself. The novel *Lucifer's Hammer* by Niven and Pournelle is probably the best of the bunch.

The larger or optically brighter asteroids like Ceres and Vesta can be easily seen with small telescopes, especially using a map of their path among background stars over weeks. Like Uranus, Neptune or many deep sky objects like nebulae or other galaxies, discerning asteroids in a telescopic field of stars can give the observer a terrific sense of achievement. The mental impression of personally spotting the real thing, be it the planet Jupiter, or Andromeda Galaxy or Epsilon Aurigae, is indelible, diminutive as they may appear compared to textbook pictures from giant instruments or the Hubble Space Telescope. These examples can actually be seen with the naked eye. The Andromeda Galaxy is marked on ancient star charts and the current dimming of Epsilon Aurigae is visible to the unaided eye. Binoculars do surprisingly well for extra power. It's worth recording what people say when they see the telescopic panorama of the Moon for the first time.

Several times in my reading and research I've had a genuine "Tunguska moment" when I really felt I was making progress and a line of reasoning was proving fruitful. I will leave it to the readers to have their own flashes of clarity in the hope of inspiring original thought. There are some 120 theories or variations published and about 25% of them are considered scientifically viable. Others deteriorate through the pseudoscientific to the ludicrous.

Firstly, a brief word on the "spaceship theory." There is not a shred of actual evidence, but absent any convincing explanation the field is open, and all sorts of fun and imaginative musings will be part of the Tunguska game until we arrive at a definitive solution. Let us note that the original spaceship hypothesis came from a science fiction plot, a piece of creative fiction from a Russian engineer, incidentally a chess master gifted with a fertile literary imagination.

I constructed a plasmoid model for my own satisfaction starting with an aberration or extreme rarity of solar origin. In simple terms, I ask whether an unusual energy body could have physically parted from the Sun and exploded on our celestial doorstep. At least two other sources, more educated than I, have proposed similar plasmoid hypotheses. Admittedly, this has no place in known astrophysics. It belongs in some "peripheral" bracket as theories go. The chapter on the Sun gives a hint of that body's awesome capabilities.

For the discerning science student, please note that I have used the terms "theory", "hypothesis" and "model" very casually. A hypothesis is a proposed and tentative suggestion and a theory is a refined account with alleged predictive powers. Strictly, the ability to disprove itself should exist or similarly the potential of detecting its own limitations should be built into a theory. A model is more a structural representation like the atmosphere of the planet Jupiter that, incidentally, we are still working on.

There is nothing man-made or artificial about the Tungus event *per se*. That there was some reluctance to publish and communicate scientific information from one source, very early in the tale, must be acknowledged to have affected the possible course of investigation. This unfortunately goes all the way back to 1908. Competition for professional offices and funding, infighting in Russian academia, non-translation of important papers (from our viewpoint) and possible incentives for national secrecy enter the equation.

Tunguska has no final authority. I do not claim to have discovered the solution to it, and it is remarkable how little progress we've made in over a century. I have described everything to the best of my ability and the faults, inaccuracies and oversights are entirely my own. Sometimes we seem to run into stranger and stranger evidence promoting increasingly bizarre theories. Somewhere out there is the truth of the Tungus event.

This book will aim to stick with science, however. Our enigma should receive a fully scientific treatment rather than a detective story investigation.

Chapter 1. Tunguska, 100 Years of Mystery

In Siberia in June, the sky stays light almost around the clock and the hardships of a long winter are forgotten. But in 1908, on the morning of 30 June, about 7:17 a.m. local time, the greatest natural explosion in recorded history occurred five miles above the remote Siberian wilderness. Whatever its cause, it devastated or damaged 830 square miles of forest and in total some 30 million trees. The blast was heard over 500 miles away. The atmospheric shock waves went round the world twice.

It was nearly two decades before any scientific investigation could take place. The heroic scientist Leonid Kulik led the first of his four expeditions in 1927. He had been previously commissioned to investigate meteorite falls in the USSR by the fledgling Soviet Academy of Sciences whilst engaged at the Mineralogical Museum of St Petersburg. After a vigorous and demanding journey toward the site of the fall, his party of explorers was dumbfounded at the scale of devastation. Yet Kulik could never locate any craters or remnant physical pieces of the Tunguska Cosmic Body itself that is presumed to have caused the explosion.

Over time he gathered many first-hand accounts of the dramatic events of that morning from the Tungus herdsmen and hunters. The descriptions included a brilliantly bright bluish white cylinder or pipe-shaped object, possibly larger and brighter than the Sun, descending for some minutes. Following this, there were hugely hot blast waves, terrifying shakings of the ground and a series of great bangs.

The engineer driving the trans-Siberian Express about 400 miles distant pulled the train to a halt, thinking some momentous earthquake was taking place or that Train 92 was derailing. He is quoted as saying he felt "a kind of strong vibration in the air" and a "roar." Arriving at the station, the train was inspected for damage or an explosion on board.

There were accounts of shaken buildings, burned wildlife and forest fires raging for days. A spurious and wholly inaccurate newspaper account suggested that passengers from the train examined the fallen object near the junction of Filimonovo but could not approach the meteorite closely because it was burning hot. This was, however, the news given on the first printed report handed to the investigator in 1921 by a colleague named Daniil Syatsky (1881–1940), who was editor of the popular-science magazine *Mirovedeniye* and a science historian. He had written a paper on the strange twilights of (Old Style date) 17 June and suggested it might be linked to a major meteor fall before the situation in Siberia became known. The article that attracted Syatsky's attention had its source in the newspaper *Sibirskaya Zhizn* (Siberian Life) of 12 July 1908 from the town of Tomsk. It was entitled "A Visitor From Heavenly Space" and was originally written by one Alexander Adrianov, obviously with major embellishment and journalistic license.

Kulik took on a determined mission after persuading the Academy, with assistance from a colleague, the accomplished geochemist Vladimir Vernadsky (1863–1945), to support a field study. He duly sought out what newspaper clippings, information and reports were available. Kulik and his companions left Petrograd, the new name for St Petersburg, on 5 September 1921 on the trans-Siberian railway, heading for Kansk.

It is likely that this preliminary visit was the first scientific probe attempted, but it ended without reaching the explosion site. Kulik's researches included distributing about 2,500 questionnaires in the environs of Kansk and Tomsk. It was established that some fiery object had passed in the sky then noisily crashed somewhere north thirteen years ago. Unfortunately, the harsh Russian winter came early in 1921, preventing any further travelling. A supposed meteorite at Filimonovo was determined to be a natural rock formation. The group investigated other Siberian meteorite falls of more normal scales as they had been directed, and Kulik returned to Petrograd in October 1922, urging that a proper investigation be organized. He had successfully established that the luminous and noisy mysterious event had come to ground further north, near the basin of the Stony

Tunguska River. It would be some time before he could lead a proper scientific study there.

Findings from other expeditions to Tunguska, if there were any, were lost over time. In fact, it is not thought that any took place; only rumors remain to suggest there might have been some attempt. No documentation of any study at the site prior to Kulik's is extant.

Innokentiy Suslov (1893–1972), as a member of the Russian Geographical Society, passed reports to the Russian Academy of Sciences who were hearing from other sources of a remarkable occurrence in Siberia. He too had published an article in *Mirovedeniye* entitled "The Search For The Great Meteorite of 1908." It was largely based on conversations with Evenki tribesmen.

Another geologist, named Sobolev, from the Krasnoyarsk Museum near Kansk, wrote of reindeer and trees being destroyed in a terrible explosion. He had spoken at length with a witness named Ilya Potapovich and others in the region. On the day of the event, Potapovich and companions had been at Teterya, some 40 miles from the fall location, and recalled the shaking of the ground and roaring sounds.

Suslov and Kulik corresponded on the subject. A police officer used the term, educatedly translated as "aerolite," in his report to the Provincial Governor. A man named Naumenko described a brilliant white elongated mass and a shining ball, irregular in shape and bigger than the Moon that became larger before loudly descending into the forest. An accomplished geologist, Sergey Obruchev, was conducting research in the region during the summer of 1924 and came to hear of a spectacular explosion and a hugely flattened forest somewhere between the Chambe and Khushmo rivers. It was wild and dangerous country too, for several reasons.

Some investigators from Tomsk and Irkutsk had visited Kansk in 1908 but apparently could make no progress uncovering the facts. The local people would not guide or assist the Russian visitors out of a traditional distrust. Obruchev heard of the damaged forest nearby whilst studying the natural features of the region for the Geological Committee but was unable to hire any willing Tungus guides to explore that area. He also on occasion met complete denials of any giant object falling from the sky. In 1925, he wrote that "the lack of time and means did not allow me to make a survey of such a large space covered by dense forest. Therefore I had to restrict my investigation to collecting new eyewitness reports." As a professional and expert in regional geology, his father Vladimir Obruchev (1863–1956) had advised on the building of the trans-Siberian railway and constructed a viable geophysical theory for the origin of gold deposits in Siberia. Many scientific papers and his five-volume geological history of Siberia were regarded as the standard reference works and won him the Lenin prize.

The whole Tunguska region is thinly populated and the Evenki people were not anxious to probe the events on any terms. Religious fear, denial and secrecy were the norm. One reads elsewhere that Suslov had successfully found some locals willing to show him the demolished forest but did not undertake the journey.

Kulik's second attempt and first real foray came in 1927. The report he delivered to the Academy of Sciences was well received on his return, and his expedition was mentioned in the Western press. Science and astronomy publications published reports and seismic and air pressure data became available from much further afield. Here it was pieced together as a recorded barometric measurement that the wave had circumnavigated the world twice.

On the second expedition, departing in April the next year, some movie footage and photographs were taken by the cinematographer Strukov from Moscow's Sovkino studio. The 25-minute film "In Search of the Tunguska Meteorite," including those shots, was produced, and an early English language dubbed copy is in the Smithsonian. Many of the original photos are preserved in an institution at Tomsk. In a serious mishap Kulik fell off a boat and nearly drowned; this was actually recorded on film. The zoologist Sytin also took part in the study.

An early description by Kulik suggested that it may have been a major meteorite in a cocoon of burning gases. As a trained mineralogist, he soon realized that this was a unique event. "A more detailed investigation of the fall region is essential," Kulik concluded, addressing the Academy after the first visit.

A telegraph from P. Sukhudaeff had been sent from Kansk to St. Petersburg's Central Seismic Commission on the very day, reporting the seismic and microbarograph oscillations. This was apparently the only immediate scientific communiqué and the sole such message sent on the day itself from the affected region. Unfortunately, due to Sukhudaeff's low rank and the geophysical experts' assumption that this region of Siberia is not particularly seismically active, the message was ignored. The Siberian craton is part of the broad geophysical heartland of a major continent, an old and stable part of the continental lithosphere. In the fullness of time, 65 meteorological stations in central Siberia were found to have recorded the aerial pressure waves.

Klykov, the postmaster at Znamenskoe, saw a "fiery streak in the sky" during what he took to be an earthquake. Other descriptions include a body "elongated and narrowing towards one end" and a "fiery dart torn from and as bright as the Sun." A peasant named Privalikhin, 15 years old at the time of the event, recalled in 1930 that he watched an elongated flying object for three minutes before it disappeared over a hill. A man named Romanoff saw a flattened ball of fire and two fiery pillars as it struck the ground. A boatman named Kokorin at the village of Zaimskaya witnessed a fiery red flame three times the size of the Sun followed by great crashes, in his words. Workers at a bell tower in the village of Nizhni-Ilimskoye saw a "fiery log flying from southeast to northwest." Another description was of a red flying ball followed by rainbow-like bands. Bryukhanov, a farmer living west of Kezhma, saw a large flame leap up from behind the forest to the north and heard sounds like gunfire. He also saw water moving *up* the river. Slightly more to the north, at Boykit, the flash of fire and loud bangs were observed to have emanated from a southeasterly direction. From the Strelka-Chunya trading post the Evenki S.I. Chuchana spoke of numerous thunderclaps and a "second sun appearing" that hurt his eyes with its brilliance. A wind flew at them, knocking them off their feet. "Now, I remember well that there was one more thunderclap but it was small and somewhere faraway — there where the sun sleeps at night."

Sitting outside at Vanavara, 37 miles to the south and the closest permanent settlement to the fall, the local resident S. Semenov was knocked off a chair whilst his house, barn and crops were damaged by high temperature blast waves. His shirt became so hot that he wanted to tear it off, he said in an oft-quoted full statement. The heat was distinctly felt in the town of Vanavara, where the object clearly passed very close at the end of its passage to Earth. Another resident, P.P. Kosolapov, heard sounds like reverberating peals of thunder receding towards the north. A pane of glass was broken in his house and due to the heat he thought the roof had somehow caught fire. From north of Lake Baikal, it seemed to have passed 80 miles west of the Kirensk and Nizhne-Karelin. The postmaster there, named Vakulin, saw a ball of fire crossing the horizon. A fiery red sphere moving horizontally was reported, the body rushing from the south to the northwest. The crashing sounds were elsewhere interpreted as cannon fire and military action. Another record holds that the sequence of noises and strong winds kept up for 15 minutes. The clear, cloudless sky and calm weather conditions are unequivocal.

It may have been 50 miles high when first seen (an estimate derived from the verbal testimony left to posterity). Such a quantified assumption will prove important to any attempt to conduct a deeper analysis.

There were many expressions of alarm at the shock and interruptions caused by the spectacle. A villager named Tropin thought there were logs being rolled into his house. Later, Golunin, Director of the Meteorological Station at Kansk, gathered accounts from people along the railway stations. At the village of Manzurka, the medical doctor Sergeyev both heard a sharp series of noises and felt a vibration in the ground, from east to north, he estimated. At the town of Khogot, the postmaster noted three explosions. As far away as Denmark, in response to the illuminated nights that followed the event, Kul, a meteorologist, speculated in his diary that a large meteorite may have landed somewhere in the world.

Arkady Voznesensky (1864–1936) was Director of the Irkutsk Magnetic and Meteorological Observatory, established in 1884. Located near southern Lake Baikal, he had made a detailed study of the seismographs and estimated that the explosion had been heard over 380,000 square miles of territory. His original interest in 1908 was purely seismic, including collecting data on minor earthquake tremors in the region two days before the event. There was no mention of atmospheric anomalies or geomagnetic storms at all. The questionnaires distributed by the observatory were not sent in enquiry of the explosion as such. He was unaware of the great phenomenon at the time as the cause of a lesser earthquake on 30 June. They were dispatched to the routine network of observers and reporters in the region, educated persons who were scientifically oriented and reliable in their assistance to the observatory's work. He eventually published a report in *Mirovedeniye* but notably not until August 1925, seventeen years after the event. His views were by no means made available at the time.

Another article appeared in the same issue, written by the younger Obruchev who had visited Tunguska 16 years after the event. He had also interviewed Potapovich et al., and written to Kulik. There had been denials and refusals to share information from the locals, as the writer observed. Suslov, as ethnographer and geographer travelling in the region, had collated some 60 reports in both Vanavara and Teterya during summer 1926 for his published article and later collaborated with Kulik on expeditions.

Voznesensky constructed the first attempted mapping of a flight path for the object and location of the fall zone. He was in possession of the most accurate information at the time. This included a good estimate of the geographical coordinates of the "impact" site and the work proved useful to Kulik's teams in the field. He suggested that a major meteorite was responsible and that a large crater awaited discovery there. A successful expedition could prove profitable, in his opinion, especially if an iron-type of meteor could be recovered.

His informative work, strictly limited to the seismic and air pressure data, was sent to the Mineralogical Museum where it reached Kulik. Voznesensky observed that it was a technological first for a meteor's apparent fall to be recorded on a seismograph but he overestimated the height of the explosion. His was the first study to show that an explosion had occurred in the air. He also expected major pieces of the landed body to be in evidence and believed that there may have been several meteors in flight and impact. He was the first scientifically-literate person to realize that a major cosmic body had descended over central eastern Siberia that day.

Exactly why he limited the information to that of earthquake activity in his report to the Seismic Committee of the Imperial Academy of Sciences in 1908 is unknown. He was curiously silent about the site of the explosion and any information on the flight of the meteor at the precise time. There was no description of curiosities in the atmosphere or magnetic storms at all. He was in possession of fuller facts concerning the events of 1908 but kept them to himself for years. Perhaps he feared professional ridicule but it is a strange academic stance to adopt, considering the corpus of quality scientific work he had assembled very early in the tale of Tunguska from an observatory fortuitously close.

The resources at Irkutsk were the best equipped in the entire region. It is possible that crucial data, at least semi technical, may have been lost regarding the projectile's passage. "Meteorite" as a term and adopted description was first applied by the Siberian newspaper reporters of the time. The reportage was rather slapdash and grossly inaccurate, as we shall learn. The data recorded on the magnetograms at Irkutsk did not surface and were not properly assessed until 1960.

Living in a village near the Angara river and some 125 miles removed from the fall site was the political exile T.N. Naumenko. He is quoted in Robstov's research with the following:

> The day was sunny and absolutely clear—not a cloud in the sky, no
> wind at all, complete silence. I was facing north. At about 7 o'clock the
> Sun was already quite high in the sky when there was a hardly audible
> sound of thunder and I quickly turned to the southeast towards the Sun.

Its rays were being crossed form the right by a broad fiery white stripe. On the left an elongated cloudy mass was flying to the north. It was even brighter than the stripe, dimmer than the Sun's disc but almost as bright as its rays. A few seconds after the first clap of thunder there was a second much louder. The flying lump was no longer visible but its tail (the stripe) was now to the left of the Sun's rays. It was getting broader than it was when on the right. Almost immediately there followed a third clap of thunder, so powerful that the earth trembled and a deafening wave resounded over the boundless Siberian taiga.

Among the replies Voznesensky gathered in the questionnaires of 1908, 17% had seen a strange flying body and 30% felt the related ground tremors. All had heard something. Kulesh, Director of the Kirensk Meteorological Observatory responded that he had heard powerful noises himself that morning and had collected first hand accounts including the visual sightings made by others. He had in the moment mistaken the thuds for gunshots from a nearby shooting range. The abrupt changes in atmospheric pressure were recorded by the barographs quite clearly, located some 300 miles east of the fall site.

A wealthy merchant named Susdalev had directed Ivan Aksenov, a Tungus native, to take a small party including some influential shamans to the area of the fall in 1910. For the purpose of preventing the exploitation of gold or diamond deposits by locals and controlling the fur trade, he deliberately engendered the rumor of an enchanted and forbidden zone, using their religious influence. He was entirely motivated by profit but this was apparently the first, if nonscientific, visit to the heart of the phenomenon by a non-Evenki person. The people avoided the area in superstitious dread thereafter. It had become an enchanted no go zone.

In 1927 Kulik's recruited guide spoke of his brother's tent 75 miles from ground zero being hurled away by a hot wind of hurricane force. Its owner was struck dumb by shock for several years following the blast. He and other hunters and fur traders had previously been interviewed by Suslov and had sketched a map of the fall zone and surroundings. It was ascertained that the affected region was three or four days' journey northward from Vanavara, depending on the season and conditions of travel. Winter miles are tougher travelling than summer miles. Suslov had provided a letter of recommendation for Kulik to the local Soviet official, arranging for him to contact Tungus chiefs, a discreet act in the political climate of post-revolutionary Russia. Suslov was Chairman of the Krasnoyarsk Committee for Assistance to Northern Peoples and politically highly influential.

Kulik's party reached Vanavara on 25 March 1927. Vanavara, originally "Anavar," means "a lucky place for hunting."

Potopovich, a key witness was initially reluctant to act as guide into the sacred land and it took a sizeable payment to him in flour and other goods to persuade him to undertake a journey there. There were many hardships and privations including shortages of supplies and funding over the seasons of dedicated work in hostile terrain and weather. Kulik and his teams did a remarkable job of surveying and observing for their severely limited resources and the occasionally dangerous circumstances surrounding them. Their markers and log cabin habitations are there in the forest to this day. We thought there were no human fatalities directly associated with the explosion. One individual named Ivan was reportedly thrown through the air, losing consciousness and sustaining a broken arm. Some bruises and many unnerving, jolting experiences exist in the verbal records. The reactions are entirely of very great shock. There are tales of whole reindeer herds being killed, buildings, tents, barns and food stores damaged with windows broken in buildings facing the direction of the Great Hollow where the TCB finally exploded. We have many descriptions of the violent noises associated. Water had been driven upstream on the rivers, watched by incredulous boatmen. Farmers and horses were thrown from their feet 400 miles away by the powerful ground shocks. The "water gushing from the earth" was clearly the sudden melting of permafrost in the zone by sudden intense heating far swifter than the usual summer thaws. A Tunguski named Onkoul grievously lost his shelter, storehouses, goods and a stated 1,500 reindeer. A farmer grasped hard at his plough so it would not be swept away. Topsoil was scattered on farmland by the hot, strong winds.

Kulik's first visit to the Stony Tunguska River basin in 1921 had aroused his scientific and professional curiosity. He wrote articles for the "Society of Lovers of World Knowledge" and the Academy's Journal on his return. Overall, he found a receptive academic audience. The collective written work of those other researchers proved important in finally obtaining the Academy's financial support to lead a party to the region. It was six years later (nineteen years after the event itself) that he first successfully penetrated the wilderness and located the apparent epicenter of the blast.

Kulik and an assistant, Oswald Gyulikh left Leningrad in February 1927 for Moscow, which had once again become the Russian capital. It is about 450 miles from Kansk to the Stony Tunguska River and an arduous journey brought them to Vanavara on 25 March. Travelling was continued entirely by foot, boat, horse and horse-drawn sleds in those days. A Tungus named Okhchen joined them as an additional guide and the expanded party crossed the Makrita River where, on April 13, they first encountered uprooted trees. They reconnoitered, climbed hills and the twin peaked mountain Shakrama and pushed on, astounded by what they were seeing. Potapovich is quoted as saying on the spot that this was where the thunder and lightning fell down.

The guides also refused to go any closer at this stage and reluctantly Kulik was forced to return to Vanavara. Wrath and divine punishment would be visited on those daring to encroach, according to the dominating superstitious belief among the party. Neither was it safe for Kulik to continue with only one companion when they had limited experience of the hazardous local terrain. One account holds that the locals were so determined to leave the area that they made the journey home in a mere two days.

On hiring other Russian hunters as guides, Kulik set out again a week later on 30 April from Vanavara. The party pushed on along the Chamba and Khushmo Rivers and made camp. Forced hacking through the forest was sometimes the only way to make progress. By 30 May they had again reached signs of devastation to the forest. At the very epicenter, the trees were stripped and severely damaged but somehow left standing at a ground zero. He described the scene as "dead forests enchanted as if in a fairy tale." He also suggested that there was "some kind of node or region of rest, due to the interference of airwaves" at that strange epicenter. It was June 1927. Biting insects were a torment. Snakes and bears were other threats.

You Tube offers a brief look at the footage of the party in mosquito-proofed clothing trudging through swamplands, sighting surveying equipment in dangerous marshland and maneuvering boats in perilously rapid cold rivers. The movie shots of the smashed forest still convey a solemn and profound impression of what had occurred nineteen years before.

Seriously short of supplies and attempting to live off the land, Kulik's team took nine days to return to Vanavara. Kulik describes that he was still trying to sort out the chaos of the impressions and the scale of the extraordinary meteorite fall. He entered in his diary:

> From our observation point no sign of forest can be seen, for everything has been devastated and burned and around the edge of the dead area the young, twenty year forest growth has moved forward furiously,

seeking sunshine and life. One has an uncanny feeling when one sees thirty giant trees snapped across like twigs and their tops hurled many yards.

The destruction to the forest lay from one horizon to another. The full devastation covered a tract of the taiga equivalent to half the area of Rhode Island.
Trees had been overthrown over a total 830 square miles and within that some
386 square miles of the forest was charred by powerful heat waves quite unlike
a normal forest fire. Rather than a regular fire consuming the lower bodies of the
woodland first from the ground upward, they had been burned uniformly, continuously and powerfully. Forest fires invariably start on the ground vegetative
litter. Later it would emerge that the pulse of heat was also brief, however powerful in effect. The charring was mostly on the sides of trees that faced the explosive
heat. There were also a few clusters of comparatively undamaged timber, oddly
suggesting a small element of selective burning from above. "Thick giant trees
snapped across like twigs," wrote Kulik. He suggested that "wave interference"
had played a part; more modern conclusions hold that the explosion(s) had not
been entirely uniform in force. Other old trees were rotted at their roots and in
danger of crashing down in some parts.

Kulik named the general area the Cauldron rather than the Southern Swamp.
From one summit he estimated an area of devastation 62 miles north to south and
25 miles east to west. The new tree growth was now nearly 20 years old and was
clearly very resilient.

A subsequent expedition conducted several draining and drilling projects by
hand in search of the huge meteorite and its assumed scattered fragments, all in
vain. There simply was no strewn field of meteorite pieces or huge buried rock
from space to be found. He had managed to circle the entire area, a remarkable
feat of surveying in itself and done entirely on foot with rudimentary equipment.
By early June, there were some serious fears for his safety back in Leningrad.
More critical suggestions were that he was taking so long because he had not
been able to find anything significant.

Sinkholes on the marshland proved to be natural formations and decisively
had not been caused by meteorite impacts. They were due to the thaw of the
permafrost; excavation proved unproductive. These thermokarst holes are shallow depressions caused by the selective thawing of the ground ice. Kulik did not
realize this at the time and was disappointed not to unearth meteorite pieces. A
1961 report described, "The thermokarstic funnels are not directly related to the
explosion of the meteorite."

Florenskiy continued, "The possible simulation of thermokarst development as a result of the fall calls for additional helological research as well as identification of the 1908 peat layer for purposes of a stratigraphic hunt for meteoric matter."

Their superficial resemblance to small craters inspired hasty judgment to identify them as such. The "Suslov Crater" most favored by the explorers as the central fall site was a complete misnomer. They also theorized that the marsh could have swallowed the impact crater and now possibly hid the massive body itself. Men were set to the arduous task of shoveling and pumping, but all this was disproven and abandoned in due course. This laborious work is also profiled in that available movie footage of the early expeditions as workers dig and pump away. The complete lack of meteorite pieces as well as the presence of a deeply buried tree stump that would have been pulped by impact weakened the meteorite hypothesis.

Kulik clung to the belief that the sinkholes were formed by meteorite impacts for so long that his scientific judgment came under criticism. He actually forbade the photographing of the notorious stump. Evgeny Krinov (1908–1984) took the picture surreptitiously. He had joined the third expedition and became Kulik's deputy in the field. Later he undertook huge research on the mystery including lobbying for an aerial survey. His book *Giant Meteorites*, eventually published in English in 1966, included a full description of the site and interviews with witnesses and was regarded as a valuable work. But Kulik believed for many years that a crater or a series of them must lie somewhere as should pieces of the great meteorite.

Reports began to reach the West in 1928, after Kulik's first full expedition. The next year the English astronomer and meteorologist C.J.P. Cave noted that the dates on six independent barographs recording significant pressure changes over England were consistent with the mysterious Siberian fireball of 1908. They had taken five hours and fifteen minutes to reach the British Isles from the fall site with four clear peaks in pressure, a very unusual reading. Instruments in Jena and Potsdam, Germany and as distant as Washington DC and Java, had also registered the pressure waves. Their average speed had been just over 700 miles an hour. Secondary waves occurred over twenty four hours subsequently. Seismic waves had been picked up beyond Irkutsk as far as Tashkent, Tbilisi and Jena.

Kulik led a second expedition in summer 1928. The Mineralogical Museum had held a meeting in February, chaired by Vernadsky, where the suggestions that the damage to the forest was caused by hurricane or conventional fire were proposed. The magnetic measurements attempted were inconclusive and that was something of a disappointment. Neither did the digging of trenches along-side holes yield any meteoric findings despite prodigious efforts. The photography and footage from this study become very familiar in Tunguska studies.

Funding was provided for further investigation and a third expedition a year later. Departing in February 1929, this was the best-equipped study yet, including drilling equipment. They arrived on 6 April 1929 and were in the field for eighteen months in all, constructing log cabins for shelter. Further surveys of the radial positions of the stricken trees placed the fall point on the south side of the swamp. The drilling and trench digging still failed to produce any pieces of meteorites or unequivocal signs of crater(s). There were no upturned strata or raised rims for a semi-buried crater wall. A giant meteor had not struck or formed the expected craterlike pits and folds of peat.

In 1933 a group of astronomers from Cambridge, England sought permission for an American research group to make the journey to central eastern Siberia and conduct a major study from the air. The Soviet government firmly denied them, on the grounds that Soviet scientists would tackle the studies of any mystery on their own territory and, after all, it is rare for a nation to voluntarily permit aerial inspections by its chief rivals. Kulik himself remained convinced that a major mass of extraterrestrial iron of great potential commercial value to the State lay beneath the Southern Swamp.

An aircraft was made available in July 1930. Kulik and a pilot named Chukhnovsky actually set off, but they were foiled by impossible weather conditions. A further attempt in 1937 was thwarted by local flooding and a hydroplane with Kulik aboard crashed whilst attempting to land on the Podkamennaya Tunguska River. Not until 1938, after interminable delays, was he able to conduct an aerial photographic survey. The large number of shots taken then proved of good utility to scientific posterity in studying the patterns of the fall zone in the dense and decimated woodlands. He and the photographer Petrov were able to assemble a remarkable photo mosaic of pictures taken aloft. At the time, he was concerned that regrowth would limit the usefulness of the pictures.

A small airstrip and a 35-mile road northwards from Vanavara leading closer to the Southern Swamp now existed, improving access. The massive butterfly-shaped pattern of the devastation was first revealed but absolutely no central crater could be seen. A reprint of the map produced by the airborne project indicates that the scorched zone extends 12–14 miles SE from the site of fall, as Kulik termed it, towards the Vanavara River. Trees were felled for a further 23 miles in the same direction. Modern satellite imagery still shows reduced forest density and irregular clearings, visible from space.

Kulik spoke and lectured on the subject very competently and was further published in the Academy's Journal and Popular Astronomy magazine. Kulik's weeks there in 1939 proved to be his last. The next approved expedition to undertake a magnetic survey in 1940 was cancelled due to the threatening military situation with Germany. Following the Great War, Russian Revolution, the fall of the Tsar, Civil War and Great Terror, once again war was impending. A natural cosmic catastrophe in the wilds of Siberia, three decades past, was quickly erased from the priority list in a world now plunged into conflict.

Despite great efforts, no single large crater, series of smaller craterlets or remnant meteorites ever came to light for the investigators, then or now. Never in the sum total of research has any such evidence been found either at the site or anywhere beneath the body's uncertain path through the sky. Given the extent of the investigations over time any such evidence would have been found by now.

There were no further expeditions until 1957–58, half a century after the event. The world had significantly changed, including Stalin's death in 1953. Khrushchev and Eisenhower were the leaders and in many ways the world had moved on. It was, as Kennedy put it in his inaugural speech, a hard and bitter peace. The cold war between East and West ensued and would stealthily remain until the early 1990s.

Eventually, new expeditions were sent to Tunguska and they were better equipped than any previous ones, by far, involving technical specialists of many disciplines. They finally disproved the existence of any craters. Their soil analyses revealed a very low concentration of magnetic and silicate dust, possibly resulting from the great explosion, possibly background in origin. More work was done on soil samples earlier returned by Kulik's earlier expeditions. K.P. Florenskiy contributed an article to the *American Sky and Telescope* magazine and led the fifth and sixth explorations in 1961–62. He had done a brief flyover of the fall zone in 1954, producing a better map. Florenskiy favored the theory that a cometary body had undergone explosive vaporization. The first post Cold War visits open to an international community of scientists took place in 1989. Over the years, many more studies have taken place.

For days after 30 June 1908, the night skies of Russia and Europe held a strange glow of permanent bright twilight, as recorded in the European newspapers. Eyewitnesses to the object itself related the "sky splitting in two" and an object from space possibly both larger and brighter than the Sun. It blazed a luminous trail as "the fire came by." The majority of descriptions recalled the huge sounds and ground tremors rather than sighting the imposing flying body. A huge dark cloud following the tongues of flame was associated with its final demise. Further away there were a few bystanders among the Evenki who "watched the tree tops get snapped off" by the power of the blast and "burning trees falling." There were Tungus people within 25 miles of the holocaust that day, possibly closer. The loss of possessions and livestock and the fearfulness of the strange occurrences cut the deepest impressions.

The newspaper *Sibirskaya* described the events as "some sort of unusual natural phenomena" from a correspondent named Kulesh, based in Kirensk, printed two weeks later. Seismometers in the locale had noted the impacts to the ground. Located 600 miles SE of the site, the meteorological station at Irkutsk recorded a magnetic storm lasting five hours. This crucial fact was not reported or appreciated at the time, yet possibly it is of major significance to determining the final explanation. Barographs in Russia and the West showed rapid pressure changes indicating atmospheric shock waves that astoundingly had encircled the globe twice. Note the magnetic anomalies reported in its wake. A Stockholm newspaper reported a "strange illumination" in the night sky as did a Dublin based publication over the next days. The *London Times* suggested it was caused by an eruption of some unknown volcano in a remote part of the world; it was reminiscent of Krakatoa and the dust lodged in the atmosphere following that volcanic eruption. There were some powerful thunder and lightning storms over London on 4 July 1908. Three meteorological stations in London and three others elsewhere in England were found to have recorded the passing pressure waves of the event. Seismic tremors were recorded as far away as Potsdam, Germany and minor ones in the United States.

The *New York Times* of 5 July relayed the news of the strange atmospheric effects seen in England. In London there were calls to raise the alarm to a huge fire, such was the sky glow over the northern part of the city. One house beheld a strange magnetic blue glow following a thunderbolt. The journal *Nature* mentioned the bright skies as did the following month's magazine of the Royal Observatory at Greenwich and *Scientific American* publications. The first illumination of the skies can be traced to 15 hours following the event.

The *Krasnoyaretz* newspaper's correspondent was reporting from Kezhma, only 130 miles from the fall zone, and managed an accurate description dated 26 July 1908. Several newspapers outside Siberia had brief columns on the tumultuous events. The bright nights were noticed on the shores of the Black Sea. On the River Volga a ship's captain was reported to have been able to see other vessels on the river two miles away after sunset. Atmospheric anomalies several days prior to the event may also have occurred, a mysterious set of circumstances in itself.

Concerning Russian newspapers, the great organ *Pravda* was first published by Trotsky in Vienna later that year and did not move to St. Petersburg until 1912, then onto Moscow in 1918. After autumn 1908, reports on the great meteorite fade from the Siberian press.

Although not a Tunguska researcher as such, the space theoretician and pioneer Konstantin Tsiolkovsky had published a seminal early paper in rocketry and astronautics in Russia five years earlier. Generally translated as "The Exploration of Cosmic Space by Means of Reaction Devices," it is probably the world's first academic treatise on rocket propulsion. His was a stringent engineering approach to the dream of spaceflight that also fascinated the literary geniuses H.G. Wells and Jules Verne. The concept of rocket staging, for example, was first proposed in his work.

Kulik was astounded to see so many trees felled and charred, evidently by shockwaves and brief but intense waves of heat. At an epicenter approximately 5 miles across were the dead but upright trees that came to be known as the "telegraph forest" or "telegraphniki." They bore a strange appearance at the very center where the explosion had taken place. The terms "zone of indifference" and "chaotic flattening" have been applied to the very early surveys. The burning had been more prolonged further away. The densely wooded Yenesei forest area had downed trees for 37 miles in one direction, it was estimated. Kulik circled the middle of the zone by plotting the directions of the fallen trees over a vast area and successfully pushed on to the very center. Kulik remained convinced it was the fall of a huge meteor that caused the carnage. By strenuous effort, he produced a very good basic map over time. Much later studies showed that nowhere in the directional studies of the fallen trees were there any significant deviations from the radial flattening noted. The influence of local relief on the timber flattening in the vicinity of the epicenter is of a nature such as to confirm that the explosion took place at some height (Florenskiy's report, p. 11).

Trees had been destroyed, strangely charred by strong heat, and thrown down like so many matchsticks on a giant scale. This was a swift destruction of a size beyond the capability of man in that era. There had been a dry spell in summer 1908 and there was a low SE to NW wind that fateful day. Studying past weather records in 1961, the Central Weather Forecasting Institute noted that "on the basis of data available to the Institute, on 30 June 1908 the site of the meteorite fall was under the influence of a zero-gradient pressure field with weak southeasterly winds at 2 to 5 meters per second."

The records suggest that there were no strong air currents at heights of a few miles and the direction of the wind was constant from the site of the meteorite fall up to 65–70º north latitude, where they deflected eastward. Overall, it was a seasonally mild and dry summer day in a region whose weather can be far more ferocious and invariably is during the notoriously harsh winters. Summer is brief at these latitudes and lacks the charm or accessibility of the Scandinavian "land of the midnight Sun." The region lies 400 miles south of the Arctic Circle and is even less populated heading northward. Maps of this region show one of the loneliest and coldest expanses on all Earth besides Antarctica.

That clutch of photos of the ravaged forest tells a remarkable story. The scorched and blackened trees are peculiar to behold and vast numbers of conifers and evergreens lie radiating outward like spokes in a wheel. Kulik's writing included that "the catastrophic impact of the leading air wave must be emphasized because according to reports from the Tungus it not only broke and felled many trees but also damned the Ogniya River, having brought down the riverside cliffs." He also entered in his diary, "ruin as far as the eye could see. What if this had been St Petersburg?"

It is a good question. Any such explosion over an urban area would have led to massive damage and loss of life plus altering the course of history. A past Guinness Book of Records notes that if the object had arrived 4 hours and 47 minutes later, then by the rotation of the Earth it would have struck the Russian imperial capital. The calculation is credited to T.R. LeMaire in his "Stones from the Stars" (1980) and concurs with a similar projection from the Russian Academician Igor Astapovich, almost to the minute. Along its approximate line of latitude to the east lie St Petersburg, Helsinki, Oslo and Bergen as major population centers.

Moscow, London or Paris are only a small part of the world's spin away. A major tsunami would have resulted if it had occurred over an ocean and major glacial melting and inundation if the explosion had occurred over an ice desert. The severe damage to coastal regions and engulfment by flooding form a nightmare scenario of potentially the very worst natural disaster in history. That fact alone stands out in the story of Tunguska yet is rarely mentioned as a close call with catastrophe.

Here is the article from the Irkutsk newspaper *Sibirskaya* from (Old Style) 2 July 1908, as quoted:

> On the morning of 17 June just after 9 AM, some sort of unusual natural phenomenon was observed in our area. In the settlement of Nizhne-Karelinsk (about 200 versts to the north of Kirensk) the peasants saw in the northwest, quite high above the horizon, some sort of body glowing with an extraordinarily intense [light] (such that you couldn't look at it) moving downwards from above over the course of 10 minutes. The body took the form of a "pipe" i.e., cylindrical. The sky was cloudless, only low on the horizon on the same side on which the luminous body was observed, there was noted a small cloud. It was hot, dry. Nearing the ground (the forest) it was as if the shining body spread out, in its place there formed an enormous puff of black smoke and there was heard an extraordinary powerful rumble (not thunder) as if from large stones or cannon fire. All the structures shook. At the same time a flame of undetermined form began to break out of the cloud.
>
> All the inhabitants of the settlement ran into the street in a panicky fear, an old woman cried, everyone thought that the end of the world had come...
>
> The writer of these lines was at the time in the forest, about 6 versts to the north of Kirensk and heard to the northwest something like cannon fire, which repeated (with interruptions) no fewer than 10 times over the course of 15 minutes. In several homes in Kirensk, the glass tinkled in the walls facing the northwest. These sounds, as has now become clear were heard in the northern Podkamennii, Chechuisk, Zavakomnii and even in Mutinskii station, about 180 versts north of Kirensk.
>
> In Kirensk at that time, several people observed in the northwest something like a fiery red sphere moving, according to the testimony of some, horizontally, but according to the testimony of others, at a steep incline.

Near Chechuisk, a peasant driving through the fields observed the same thing in the northwest.

Near Kirensk, in the village of Voronina, the peasants saw a fiery sphere falling to the southwest of them i.e. to the side opposite the one where Nizhne-Karelinsk is situated.

The phenomenon has aroused a mass of interpretations. Some say that it was an enormous meteorite, others that it was ball lightning (or a whole series of them.)

Author's note: A verst is an obsolete Russian and Eastern European unit of length equal to 3,500 feet. It is made up of 1500 arshin units, equal to 28 English inches each. The German werst and Finnish virsta are equivalent and the best Russian transliteration is "vehrsta" with a plural "vehrsty." The shaky translations themselves cannot be airbrushed now. They do vary depending on exact source.

CHAPTER 2. METEORS, CHAPTER OF ASTRONOMY

The Krakatoa volcanic eruption of 1883 was another large scale natural catastrophe and one that claimed some 36,000 lives a quarter of a century earlier.
Traces of the volcanic dust and ashes unleashed in the event hung in the atmosphere for fully two years afterwards. By contrast, the Siberian conflagration cost
not a single life directly, as far as anyone could tell at the time.

By most approximations of the speed and obviously "height" of a body entering from space, the object easily could have covered the extra distance to hit a
major city. A cataclysmic natural disaster in Europe in 1908 killing hundreds of
thousands could have obviated the First World War, for example, and directed
human energies to researching threats common to all on the planet, potential
risks from the far reaches of the universe. Instead, the event nearly slipped into
oblivion. It remained rationally or empirically unstudied for nearly two decades
and it is entirely to the scientific credit of one man and his collaborators that the
truth of the tale began to unfold.

Past events involving major collisions with the Earth most prominently include the death of the dinosaurs 65 million years ago. Their extinction is thought
to have followed a massive asteroid impact to the (present) Yucatan Peninsula
in Mexico and the sea lying north of it, followed by the major environmental
shake up it caused in the biosphere. The semi-submerged and entirely buried
Chicxulub crater has been identified as the exact location of impact, 110 miles in
diameter and named after the nearby town. In the biological great chain of events

this may have cleared the evolutionary way for the rise of mammals, including humans. The great lizards were indeed dominant and very diverse before their abrupt demise. It has been deduced that a massive global cooling followed that event due to the sharp drop in atmospheric transparency. The material thrown up by the impact dramatically diminished the heat and light of the Sun on a worldwide scale. The dinosaurs have very few comparative living descendents, so abrupt was their demise. Birds are taken as their living heirs rather than the lizards roaming the Earth today. And this was not the only mass extinction in Earth history possibly attributable to an asteroid strike. The Permian extinction 251 million years past was probably the most exacting of the five Great Dyings that have occurred in the last 500 million years.

In very early Earth history, comets may possibly have played a major role in developing the great liquid oceans we know. The chemical load that makes them up may have been imported from space rather than developed in the primeval soup of Earth's atmosphere. Comets in huge numbers may have been instrumental in biogenesis itself.

The paths of evolution are by no means clear but there was a massive upheaval those 65 million years ago. One would think that considerable research would be directed to studying the origins of such game-changing events.

Mentally grasping the geological timescales forms a neat middleground of human and cosmic history. The "past" is a relative term. Human civilization as we know it dates back to at least 5,500 years ago in the Near East. Technology as used to improve the human condition may have been applied as early as Neolithic Man or at the latest the Bronze Age, if by technology we mean the application of fire and the use of crude tools. Civilization is better described as a social and unavoidably political set of conditions requiring a distribution of labor and social interaction. A complete and full-blown form of written language is one of civilization's more cerebral attributes.

Returning to the natural world that is our stage, the last 10,000 years has been termed the Holocene epoch. This follows the Pleistocene epoch that began some 2.5 million years ago. Those 2.5 million years also marked the commencement of the Quaternary period, preceded by the Tertiary period. These two periods comprise the whole Cenozoic era. The abrupt and eventful transition between the Tertiary and preceding Cretaceous period is termed the K/T boundary, occurring 65.5 +/- 0.3 Ma (mega annum or million years ago). Cretaceous is Latin for "chalky" and its German translation "kreide" gives rise to the abbreviation "K".

Geological time makes a stupendous chart, and one simple conclusion it suggests is that life commenced relatively early in Earth history.

Man is very debatably 2 million years old as a species, possibly older. This raises the most profound of introspective questions relating to the human condition. The Earth is an estimated 4.6 billion years old and the greater cosmos is said to go back 13.7 billion years. According to this theory, the world has only existed for approximately one third of the time that has elapsed since the Big Bang. The Sun is a little older because the proto-Sun was ignited by undergoing a hydrogen flash before the full formation of the planets.

How life began is of course a matter for philosophers as much as for scientists to debate. As for how and when life has been partially erased or seriously damaged, one suggestion in the evidence of the Ordovician mass extinction 450 million years ago is that it was caused by a gamma ray burst, by lethal radiation from a relatively nearby (on the stellar scale) star. Note that there is no clear direct evidence of this, although if a major GRB did erupt in the extended stellar neighborhood that would certainly be devastating. GRBs are generally considered to have their place in the distant outer cosmos and, of course the formatively remote past, where far away is equal to long ago.

Contemporary studies of Near Earth Objects and "Earthgrazer" asteroids led Professor Richard P. Binzel of MIT to pioneer a scale by which to assess the severity of the impact hazard. This became known as the Torino Scale. The first version entitled *A Near-Earth Object Hazard Index* was presented at a United Nations conference in 1995. A revised version was adopted at an international meeting on NEO's held in Torino (Turin) in 1999. An object is assigned a 0-10 value based on the collision probability and kinetic energy is expressed in megatons of TNT. We cannot retrospectively assign the Tungus event a reliable figure on the scale.

Weathering, erosion and natural vegetative growth act as dynamic and efficient forces on Earth. The sites of many past collisions are no longer clear and distinct on the landscape, especially the geologically older ones. Instead they are progressively obliterated as features. There are an estimated 160 large and recognizable impacts extant on the Earth's surface today. South Africa's Vredefort Dome is the very largest and is a World Heritage site. It is immensely old, perhaps 2 billion years, and has been eroded down to an eight-mile diameter over time. Similarly, the Manicouagan Crater in Quebec formed 214 million years ago and the original 60 mile diameter has been scaled down to a mere 45 miles by the processes of erosion and glaciations.

But in living memory, Tunguska remains the only time that the Earth was clearly struck by a sizeable cosmic body. (That it may have been an indirect impact is profiled in the final chapter.)

Less devastating falls exist in both reliable and sketchier allegorical records from the past and some colorful legends. Specifically, an ancient Finnish poem called the "Kalevala" describes an event of fire from the sky.

There were allegedly ten deaths by a meteor collision with some chariots in China in 616 BCE. Elsewhere in the ancient world, the Roman writer Pliny the Elder (23–79 CE) was a naturalist himself and referred to something as big as a cartwheel impacting Thrace (located in SE Europe, north of Greece) in 476 BCE.

Evenki legends themselves portray a visiting and vengeful thunder god active in the Tunguska region. During November 1662 (Old Style) in another Russian territory, the village of Novy Ergi, another major fall was recorded. The local priest's surviving account bears a resemblance to a scaled down but distinct impact event including shaking ground, fire from the sky, and smoke. Hot stones of different sizes are said to have fallen in the fields and streets and to have bored into the ground; meteorites, perhaps.

It was not until the early 19th century that the idea stones could fall from the sky was even accepted as a possibility. Meteoroids in space entering the atmosphere as meteors then falling to the ground to become meteorites were not yet proven as scientific facts and occasional actual occurrences. Thomas Jefferson famously could not accept the notion on hearing a report of such a fall event from back east in Connecticut in 1807. The Yale professor of chemistry Benjamin Silliman had investigated the incident and believed it was of cosmic origin.

A major meteorite shower fell over L'Aigle, France on 26 April 1803, as investigated by Jean B. Biot. This was a turning point in showing that meteors were genuinely extraterrestrial in source. In modern analysis, they were ordinary chondrites of type L6, a low iron group of specific texture and mineralogy. France had earlier witnessed a fall in 1768 that had been accorded a more scientific and less superstitious reception. The Italian professor Ambroglio Sodani suggested that the Sienna stones of 1794 had fallen from the sky and had not been windswept from a Vesuvius volcanic eruption very shortly earlier. He recognized the meteorites from previous studies of other small rocks supposedly acquired from the sky. The work "On the Origin of Iron Masses," published earlier the same year by Ernst Chladni (1756–1827) suggested not only that fireballs form in the atmospheric passage of meteors but also they originate in "cosmic space" as planetary debris. As theoretical astronomy goes, this was brilliantly perceptive for the time. He was ridiculed to the extent of one critic suggesting that the author had been struck on the head by such a stone. He went on to become one of the founders of the science of acoustics.

Nevil Story-Maskelyne (1823–1911) devised the first classification system for meteorites in the 1860s. He divided them into three types by basic composition: stones, irons and stony-irons. Irons are generally fairly pure metallic nickel-iron. This has proven a good basic categorization with huge advances in analytical chemistry following on, the modern terms being aerolites, siderites and meso-siderites respectively. The differentiated meteorites are formerly part of a larger body where separation according to density took place whilst still molten. Un-differentiated meteorites derive from smaller bodies that cooled and solidified too quickly for differentiation to take place. More advanced electron microscopy techniques provided a significant discovery with an individual meteorite re-trieved from the Antarctic — where an alien stone tends to be more conspicuous than one buried in a forest, mountain or sea floor.

The science dictionary describes a meteor as a small body of matter from outer space that enters the Earth's atmosphere becoming incandescent as a result of friction and appearing as a tiny streak of light. The definition of a meteoroid by the International Astronomical Union is "a solid object moving in interplanetary space of a size considerably smaller than an asteroid and considerably larger than an atom." The Royal Astronomical Society suggests using the term meteoroid for objects having dimensions between 100 microns and 10 meters. (A micron is a millionth of a meter, or a micrometer, and a typical red blood cell is about 7 microns in scale.) Very small particles are appropriately termed micrometeoroids.

The International Astronomical Union defines a fireball as a meteor brighter than any of the planets at visual magnitude -4 or greater on sighting. Two distinct mechanisms are at work, the burning of the solid body itself and the incandescence of the atmospheric gases surrounding it. Appreciable drag kicks in at about the height of 90 miles and below that considerable resistance increases. This strongly applies to satellites deorbiting and plunging to destruction. With meteors, kinetic energy is converted to heat and a temperature in the region of 1,600º C is momentarily created along their paths. The rock faces liquefy and the ablated material vaporizes. It is actually the ionized air that releases light in re-capturing electrons to form the fleeting trails of light. The colors associated with fireballs derive from the chemical composition of the meteor's material combining with the surrounding air. This is clearly understood for the tiny descending natural bodies from space that occur every day and must apply to the behavior of all natural bodies including the TCB. How the assumed ablation process could avoid leaving any trace of a material trail on the final passage downward is unexplained. We can only conclude that there was no significant ablation leading up to the final explosion.

Isaac Newton (1642–1727) held that space was empty, inside the Moon's orbit, apart from the all-pervasive "ether." That view too was discarded and rendered obsolete by the advance of science in the later 19th century. Disposing of the idea of a luminiferous ether (or light-bearing aether) had to wait until the Michelson Morley experiment 1887, termed the most famous and informative failed experiment of all time. Light does not require a medium to propagate, unlike sound. It is a form of electromagnetic radiation that shows a particle-wave duality. There is plenty of interstellar gas visible telescopically and on very clear night one can see the zodiacal light with a slightly greater concentration of matter lying in the plane of the solar system and illuminated by the Sun. It can be quite noticeable in dawn and dusk skies standing up from the horizon. Meteoric dust and debris from the ceaseless orbit of comets add to it, but the majority is probably illuminated primeval material on a tiny individual scale, left over from the stuff of the proto planets. It is not the same as the fundamental and overly philosophical "medium" encompassing all matter, although the term "ether" is still loosely applied.

The Leonid meteor stream made a particularly active appearance in 1833 with a grand display of natural fireworks, probably the best in the history of properly recorded showers. Abraham Lincoln was one casual witness to that annual storm as it hit a particularly high point that year, and he was suitably impressed. Denison Olmstead (1791–1859) conducted a study and concluded that it was decisively extraterrestrial and not atmospheric in origin. In Germany, Heinrich Olbers (1758–1840) predicted a return of major Leonids in 1867 among other contributions to astronomy. Some refinement by Hubert Newton (1830-1986), including historical research, adjusted the date to 1866 and this proved correct. He was a leading authority on comets and meteors of his era. The sight of many blazing meteors radiating from a single point in Leo was clear to all onlookers during that memorable performance. This reliable shower occurs in November every year and has what have been estimated to be the fastest meteors of any regular showers arriving at about 44 miles a second. The lowest speeds of entry are probably about 5 miles a second. The Italian observer Giovanni Schiaparelli (1835–1910) decisively connected the Leonids to Comet Tempel-Tuttle by proving that their orbits coincide. This strongly suggested that meteor showers could be the trails of comets and was another major step in understanding these phenomena.

Mostly, what is causing those celestial flashes and streaks are tiny stones and pebbles or rocky particles as small as grains in size. Generally, their visible careers last mere seconds. The larger, denser and predominantly iron ones may survive to the surface partly intact and are common sights in textbook photos and museums. They can fragment in flight or split up, leave brief smoky trails in the lower layers of the air or explode spectacularly. The mechanism by which acoustic waves are emitted and sound reaches observer's ears is still imperfectly understood. One generally gets to see one or two during a night's observing under a clear, preferably moonless celestial dome. By happenstance, August is the best month for meteor observing including the Perseids and six other reliable minor radiants. They are generally named after the constellation or positional star of apparent origin. Microscopically and chemically, meteors have been intensely scrutinized over the years.

The physics of atmospheric entry for a very large body is more academic and some imposing computer simulations are available. These case studies are theoretical and satellite observation networks (designed principally for militarily defensive purposes) enable the calculation of mass, density, volume, speed and angle of approach, the most important parameters. Most dedicated backyard astronomers have seen splitting up, burning up and exploding on the small scale from time to time. Regular, daily meteors become visible at altitudes of 40–75 miles and disintegrate 30–60 miles up, glowing for a blip or a few seconds. Their ionization trails may last up to 45 minutes. In the upper atmosphere they are constantly occurring. Smoke trails can sometimes be left for long minutes in the lower levels of the air. Watching one of them slowly twist, fade and dissipate in the wind is one of those a rare and memorable treats in practical astronomy.

Probably the greatest single meteorite fall on stringent record occurred near Kirin, China, where over 4,000 pounds of material fell in approximately 100 pieces over 260 square miles in 1976. The largest known body resting on Earth is at Hoba West, near Grootfontein in Namibia, weighing in at an estimated 55 metric tons. A necklace made of such grounded shooting stars was identified among ancient Egyptian artifacts. Some ancient jeweler or craftsman obviously had an eye to pick them out and set to work. Alternatively, and sadly for the book of lost knowledge and missed opportunity, a major fall that occurred in the Brazilian

Amazon in 1930 was never thoroughly investigated. The crater is undoubtedly reclaimed by jungle by now but the impact was big enough to shake the ground and create forest fires as it formed.

Up to the time of the Apollo Moon missions, meteorites were the only material formed beyond the Earth available for study. The Peekskill meteorite in West Virginia in 1992 flashed a widely seen 40-second blaze during over a track of some 500 miles. Recorded on at least sixteen camcorders, it did serious damage to a car when it finally came to ground. At one count the Hubble Space Telescope, in Earth orbit, had sustained 572 tiny craters and chipped zones from micrometeorite impacts over the years. Back on the ground, security cameras have caught the passages of meteors in recent times.

The Martian meteorite labeled ALH 84001 may, very debatably, contain the remains of fossilized bacteria from the distant past on Mars. The announcement from the NASA team in 1996 caused quite a stir. President Clinton fielded questions from the press on the potentially very important discovery. The organic chemicals known as polycyclic aromatic hydrocarbons (PAHs) present in the meteorite, originally found in the Antarctic's Allan Hills (hence its labeling ALH), may have been produced by long dead life forms in the process of decay. Inorganic processes could also have formed them, and here the discussion is launched. Amino acids, the linear chains that comprise proteins so vital to biochemistry, have been identified in other meteorites. The whole scientific matter surrounding ALH 84001 remains unsettled as does the much broader possibility that life may have begun on Mars but was snuffed out by major environmental changes there. The specific specimen formed an estimated 4.5 billion years ago and was shocked or broken 3.9 billion years in the past. These are major timescales in local cosmic history, bearing in mind that the solar system has about a 5-billion-year pedigree.

Exobiology as a science has a long and distinguished career ahead. The sands of Mars could provide material for research, not to mention any other possible samples such as, in the best case imaginable, fossils. Any geological feature like a dried up lakebed could provide a gold mine of interesting deposits there. That the planet had water at one time has been long proposed, based on evidence like the alluvial features and desiccated basins that have long been apparent from Mars orbiters and surface probe's wanderings. Life on Earth is said to have first arisen in the sea but here, unlike Mars, the hydrosphere and atmosphere have been strongly retained. The atmosphere of Mars is thin and that of the Moon is a good vacuum with some hydrogen atoms present.

There are, fortuitously, 34 meteorites here on Earth (at the last count) that display a ratio between the isotopes of several inert gases identical to the ratio found on Mars by the Viking landers. We thus conclude that these meteorites are Martian in origin but no one can be quite certain about the impacts and paths of their journeys here. ALH 84001 was located, among others, in Antarctica having been violently lifted from the surface of Mars an estimated 14–18 million years ago. We deduce that it originally crystallized from molten rock as long ago as 4.5 billion years and impacted Earth about 13,000 years ago. It was conspicuous, lying in the white ice, as are others. Exactly where, when or how the impact event occurred on Mars, blasting it into space, is unknowable. It was found in late 1984 and the debated traces of past life were discovered by electron microscope in August 1996. There are also 15 known genuine lunar meteorites. It is not impossible that small pieces of planet Earth are similarly at rest on the surface of Mars or the Moon, although Earth's gravitational field is significantly higher and less likely to give up chunks of material to speeds above escape velocity.

Microorganisms that have been dead for millions of years are a step in the right exobiological direction, if we are correctly interpreting ALH 84001. However, microfossils from antiquity are not exactly what most of us have in mind when we dream about discovering life in the universe. People have even expressed disappointment that the chemical remains of deceased bugs from the deep past is the most significant finding so far as to what purports to be an active search for life in the universe. The idea that there could be an exchange of radio signals from creatures from other stars cannot be ruled out, and thus we look to the SETI initiative.

Graphic signs of the Late Heavy Bombardment are elsewhere in the inner solar system such as the surface of Mercury. The most reliable figures say that the onslaught ended 3.9 billion years ago with trace evidence of life dating from 3.83 billion years.

A mere glance at the surface of the Moon allows a view of the crater-strewn and "magnificent desolation" as Aldrin termed. (Buzz Aldrin is jointly credited with humanity's first visit to the surface in person.) The craters, maria and mountains have accumulated over 3–4 billion years with extremely little change going on since. Hence, the lunar surface is old even by the timescale of the solar system and virtually unaltered over the ages by the active processes so familiar here. It has "a stark beauty all its own" yet is very low in action in this epoch. The two-week-long passage of the Sun across the lunar sky and the equal and opposite phases of the stationary, spinning Earth are the only visible movements apart from the orbit of the Sun whilst accompanying the Earth. The only action there is celestial in sight and the only regular changes are the temperature swings of day and night.

The Late Heavy Bombardment, a series of great impacts also known as the Lunar Cataclysm, occurred between 3.8 and 4.1 billion years ago. The visible face of the Moon strongly retains the evidence, preserved intact from earlier cosmic history. It is like a permanent scoreboard of impacts from space and eruptions from within of great sheets of magma. Those are comparatively less old because the mare basins are far less crater populated. One can make some sense of it with the simplest observation and maps.

The lunar far side is not so covered in the mare basins and vast lava flows that presumably followed the giant impacts, but we are entirely dependent on moon probes to give us any views there. The first pictures of the hidden hemisphere from the Soviet Luna 3 in 1959 made a very deep impression. Concerning the Lunar Cataclysm, one may speculate that one maximally intense bombardment took place over a very short period of time. The process was so brief it apparently struck only one side of the Moon. The Moon was closer to Earth and speedier on its path back then and it continues to this day to very slowly expand its orbit. At what point it settled down to captured rotation is uncertain but clearly this comprises another artifact from yesteryear. It spins on its axis almost in the same it takes to revolve around the Earth. Currently, its mean distance is just under a quarter of a million miles and its speed around the Earth half a mile a second.

Its face is noticeably different on the far, unseen side from Earth although it does indeed receive as much sunlight as the lunar nearside as the Moon revolves slowly on its axis in the course of a month. The term "farside" is more accurate than "dark side." Consult a lunar map or if available a full globe and the contrasts are quite pronounced. A slightly thicker crust on the far side and a mildly offset small core are also the geological case with the Moon.

A controversy about the lunar craters being of impact or volcanic origin raged for decades in the astronomical community. The 1949 book by Ralph Baldwin, *The Face of the Moon*, was an important early work coming down strongly on the side of impact over volcanism. This proved to be the correct although lunar outgassing with miniature volcanism still occurs. This gives rise to "transient lunar phenomena" like very tiny colored hazes in the crater Aristarchus. Mascons, slight gravitational anomalies, exist on the Moon, where the crust has been so compressed that the mass concentration of surface material creates a mildly stronger and regional gravitational pull.

Films shown to the public depicting the Moon buggy drives of the Apollo astronauts give the effect of a very close horizon from ground level and present serious difficulty in mentally gauging the distances of other surface features. It is a smaller world with no shades or hues of color that the mind is used to in judging perspective. Hills look strangely close.

The accumulation of meteoric dust on the surface always gives me a sense of sheer awe for the immense time scales required for it to passively build up. Neil Armstrong remarked that it was "almost like a powder" as he stepped onto the lunar surface for the first time. It has been described as like walking on snow. Before leaving the Moon, let us note that the best or least ragged theory for our companion's origin is termed "collisional ejection." Very long ago, some big body gave the proto Earth such a blow that a large stream of material was ejected to efficiently form a large permanent satellite. It settled relatively quickly with enough material to assume a quarter of the diameter of its parent body. Exactly how swiftly is uncertain.

This was quite early in their entwined histories. The Moon has therefore been there a long time but not quite as long as all the time there is since the formation of our world. There is a rationale apart from space science that reveals this. The adaptation of life to the tides in a major biological mechanism and obviously has been a factor over enormous time. It is the Moon that turns the tides, gently, gently away...

With the exception of Pluto and Charon, this is the closest comparative size of a primary and satellite pair in the entire solar system. As a proven Kuiper Belt Object, there is wisdom in relegating Pluto and its collective three moons from that "classical" to "dwarf" designation. There has been some acrimonious debate on this topic. The Moon is the fifth largest in the entire planetary family of satellites. Jupiter has three and Saturn one satellite physically larger than the Earth's Moon with its 2,159-mile diameter. In the classical planets, then, the Moon is by far the largest satellite of all compared to its parent body.

The hypothesis of the Moon forming from a very big impact of something with the proto Earth is also inelegantly called the "big splat theory." By extension it can be made to cover even more ground in the history of the inner planetary system. It is the least poor fit of a theory of formation and we are stuck with it for now. Formally, the competitor theories of binary accretion, intact capture, or fission as a single body are all dynamically or geologically untenable.

For example, if it was the planet Mercury that collided with us way back then, the anomalously large core of the closest world to the Sun is expounded by the crust and mantle becoming dissipated in the powerful crash. This could mean that the Moon was partly made of the planet Mercury's original crust.

Alternatively, the slow and retrograde rotation of Venus resulted from the primeval collision that sent a major plume of matter into orbit around the Earth. The blow both reversed and retarded the spin of Venus. The Moon retains 1/81 of the mass of the Earth in ratio of the material present in each body. It is broadly true that the youngest visible features and rocks on the Moon are older than the oldest surviving on Earth. All lines of enquiry conclude that the Moon has been there a long time, such as its captured rotation that clearly took eons to settle down. There are other cases elsewhere in the solar system of that tidally locked or captured rotation.

The least poor explanation seems to be "collisional ejection" and at great distance in time, too. The Earth and Moon did not form side by side from the same primordial cloud of material because this is geologically unsupported. The Moon is slowly moving away and was never further apart to join Earth intact, spiraling inward from an independent origin. The historical evidence is clear there. Lastly, if it tore away as a single entity, then dynamically it would have kept going and not formed a close satellite. There is a joke that, having considered every viable hypothesis, we deduce that the Moon cannot exist.

The Greek philosopher Aristotle (384–322 BCE) reasoned that meteors were purely atmospheric in origin, synonymous with thunder, lightning and rain, as were comets. "Meteorology" as a term and science is primarily the study of the atmosphere and weather patterns to this day, extending to climatology. The root Greek word "meteorors" means "high in the air." The Greek "kome" or "hairy" refers to the following tail and is the origin of the word "comet" as "kometes" in the sense of a "hairy star." The Chinese used the term "guest" or "broom" stars relating to comets. Aristotle therefore placed them in the closer spheres surrounding the Earth in his prodigious studies of the natural world. For once, astronomers have to be content with something lesser, the term "meteorics." for the study of meteors. The adventures of the "Phusikoi" in the Pre-Socratic ancient Hellenic world of Greek Philosophy set the very tenor of rational enquiry and our intellectual debt to them is formative. Much of Aristotle's thinking became formalized and orthodox by definition in the stuttering and religiously dominated scholastic life of medieval Europe.

Shortly before the epochal application of the telescope by Galileo (1564–1642) during 1609, the Scandinavian nobleman Tycho Brahe (1546–1601) had performed meticulous astronomical observations over decades from his fief like island observatory. This included successfully reasoning that the comet of 1577 lay beyond the distance and sphere of influence of the Moon. It showed less parallax, actually none at all compared to the Moon, whose place among the stars at a given time is slightly but visibly different from vantage points hundreds of miles apart. Most significantly, real changes like a passing comet and the chance nova of 1572 in Cassiopeia do occur in the celestial firmament. This challenged the very foundation of the received Aristotelian doctrine that the heavens are perfect and changeless. It was also the death knell of the notion of crystalline celestial spheres propelling the planets in awkward epicycles around the stationary Earth. Our earthly stage no longer rested at the center of creation. The orthodox Ptolemaic system was crumbling fast. The Copernican revolution was gathering force.

Historically, then, Tycho's one-man golden age of detailed observations and angular measurements (he carefully recorded two decades of the motion of Mars) was shortly before the application of simple telescopes in the early 17th century. At great length and applied effort, his sometime student Johannes Kepler (1571–1630) derived the laws of planetary motion from the Mars material he literally inherited from the aloof and cantankerous master.

Through Galileo's highly publicized work and not without serious resistance from Church authorities, the rugged face of the Moon was being revealed in detail as were the moons of Jupiter and the phases of Venus. In England, Thomas Harriot had been watching the Moon six months earlier with a small telescope but chose not to publish any results, drawings or theories. There is, therefore, another pioneer whose knowledge was not propagated into any public domain by his own choice.

On the subject of expanding horizons, mental and geographic, here was something of the spirit of the Renaissance. Galileo and Bruno were to pay a high personal price for their endeavors and contributions to science and the intellectual. Both severely collided with the rigid control of the 17th century Catholic Church in matters philosophical and doctrinal. Observation, experiment and the scien-

tific method were to replace dogma and religious authority. Reason would triumph over revelation. Both the broad sweep and precise details of the workings
of nature were being newly explored on bold scientific footings. Kepler worked
doggedly on planetary motion for years on end. Later, Newton was a veritable
blaze of genius on motion, gravitation, light and many other areas. Bertrand Russell once pointed out that any idea that the Earth is the hub of all things and Man
the central character is a dangerous illusion, best cured by a little astronomy.

CHAPTER 3. TUNGUSKA GROUND ZERO

When consulting the geographical coordinates of ground zero above where
the fireball erupted (Latitude 60º 53' 9" N. Longitude 101º 53' 40" E) one is struck
by the remoteness of the area. It lies about 37 miles north of Vanavara, which was
then a small trading station and remains the nearest town. The Arctic Circle is
approximately 400 miles north of the settlement.

Tunguska is the name of a general region with no strict borders located in
central eastern Siberia. It refers to any of three rivers, all tributaries of the Yenisei.
The Lower (Nizhnaya) Tunguska is 1,670 miles in length, the Stony (Podkamennaya) Tunguska is 930 miles long, and the Upper (Verkhnyaya) Tunguska is the
lower course of the Angara River. A 1200-square-mile nature reserve was established in October 1996 by the Russian government. The formal name is the Tungussky Zapovednik and according to the "Ecotourism in Central Siberia" website
it covers 296,700 hectares.

Geographically, the explosion took place in the region of two rivers in Siberia, the Lower Tunguska and Stony Tunguska (Podkamennaya) that flow west
through the forested, sparsely populated Tunguska basin into the Yenesei River.
This great river system rises in Mongolia and drains a major part of central Siberia
following a northerly course to the Kara Sea and Arctic Ocean. When looking at a
general map of Asia or a globe to locate the fall site, the blue curve of Lake Baikal
is the only clear marker. It is somewhere north of this area where the TCB prob-

ably was first seen. Unfortunately, not a single first or second hand observation or folk yarn has ever emerged from the entire geographical zone.

The bearing of the epicenter was well approximated by Kulik both on the world grid and local maps. Subsequent work measuring the azimuths of thousands of fallen trees rendered a precise set of coordinates that bore him out (Fast, et al., 1983). Further seismic wave research about the same time showed that the explosion had occurred between 0 hrs 13 mins 30 secs and 0 hrs 13 mins 40 secs GMT. The textbook references to Tunguska could be updated here as could the over-confidence of cometary theory normally described there.

It is estimated that a mere 5% of the Earth's surface could sustain such an impact without significant loss of life or adverse environmental effects, at least in the short term. With the Earth's surface over 70% water and 10% permanently icebound, only a surprisingly small amount of the Earth is comfortably available for human habitation. The Pacific Ocean alone covers 30% of the surface.

As a map exercise, those coordinates compare favorably with Voznesensky's original estimate of Latitude 60° 16' N and Longitude 103° 06' E. His observatory was instituted in 1884 and commenced with meteorological and magnetic measurements in 1886. Voznesensky had a rounded education in physical science; he became the director in 1895. The instrumentation and range of equipment was good, practically state of the art for the time.

The name for "Tungus" is the Russian, originally Turkic, term for the Evenki people of northern Siberia. The name of the greater region was the tribal "Sibilla" as derived from the Turkic nomads later assimilated to Siberian Taters. The modern usage of the word in Russian language appears following the conquest of the Siberia Khanate in the late 16th century. In entirety, Siberia is half again the scale of the continental U.S. and 4½ million square miles in extent. It has no formal boundaries as a geographical landmass comprising most of northern Asia. It ranges from east of the Ural Mountains including the Russian Far East to the Pacific Ocean. From the Arctic Ocean in the north it reaches southward to the political borders of Mongolia and China. It is a vast wilderness on a par with the hot and cold deserts of Earth, and is proverbially the greatest area of desolation of all. Further north the biome shifts from taiga to alpine tundra where the tree growth is restricted by permanent low temperatures. Such an environment is equaled in northern Canada and Alaska.

"Evenki" is a 17th-century Russian term for the Tungus tribes who are among Siberia's oldest inhabitants. It originated in a term supposed to mean "he who runs swifter than a reindeer." The nearly one-million square mile area incidentally containing the impact zone is nowadays named Krasnoyarsk Krai. Politically and administratively, it is Russia's second largest federal subject. Modern Siberia consists of twelve federal districts. The living Evenki population is given as 30,000.

There is still major uncertainty in the force and mega tonnage of the blast itself. This megaton unit of explosive power is equivalent to a million tons of regular TNT and used in describing atomic explosions and hypothetically, the Torino Scale. A kiloton is equivalent to a thousand pounds of TNT. Most other parameters like the path, speed, mass; composition, density and tensile strength of the approaching body cannot be precisely defined either. Estimates run to a thousand times the explosive force of the Hiroshima bomb at 10–15 or as high as 30 megatons. The Hiroshima bomb was equivalent to 13-15 kilotons. The Tungus Event was therefore one third of the scale of the Russian Tsar Bomba of 1961, the largest nuclear device ever exploded, or roughly equivalent to the American Castle Bravo test of 1954. These approximate figures were originally based on the recordings of the barographs in Western Europe, developed as a meteorological instrument in 1903. The calculations were made much later. Subsequent and more sophisticated computer simulations of the devastation tend to a higher value for the blast. The thousand Hiroshimas tag may itself prove an underestimate. There is, of course, healthy academic dissent to that statement. The lack of data seriously cramps our speculations.

In the words of the party possibly closest to the event that morning, the Potapovich family:

> Early in the morning when everyone was asleep in the tent it was blown up in the air, together with the occupants. When they fell back to Earth, the whole family suffered slight bruises but Akulina and Ivan actually lost consciousness. When they regained consciousness they heard a great deal of noise and saw the forest blazing around them and much of it devastated.

Another witness at Vanavara stated:

> I saw the sky in the north opened and fire poured out. The fire was brighter than the Sun. we were terrified but the sky closed again and immediately afterward, bangs like gunshots were heard. We thought stones

were falling, I ran with my head down and covered because I was afraid that stones may fall on it.

Whilst in the town of Kursk it was reported:

> A ball of fire appeared in the sky. As it approached the ground it took on a flattened shape.

And from Kansk an individual named Sarychev said that:

> ...suddenly I heard at first a noise, as from the wings of a frightened bird coming from the south toward the east toward the village of Antsyr and a wave like a ripple went up the river in the direction of the current. After that there followed one sharp crash and after it hollow seemingly subterranean rumbles. The crash was so strong that one of the workers, Yegor Stepanovich Vlasov (he is now dead) fell into the water. With the noise, there appeared in the air a radiance, circular in form with nearly half the dimensions of the full moon with a bluish hue flying quickly in the direction from Filimonova toward Irkutsk. After the radiance there remained behind a trace in the form of a blue gray streak stretching along almost the whole way then gradually disappearing from its end. The radiance, without breaking up vanished behind the mountain. I didn't notice the duration of the phenomenon but it was very brief. The weather was completely clear and calm.

From the Ilimsk meteorological station the observer Polyuzhinskii wrote four days later in a letter:

> On 17 June 1908, at 8 hours and 30 minutes in the morning there was heard a powerful noise and a sound resembling strong thunder and cannon shots, following one after another (like small shot) probably from the meteor (aerolite) flying through.

In a report a month later he conceded that the timings were in error but he had been informed of "a flying star with a fiery tail" and a "flying sphere" whose witnesses became frightened and immediately ran home from the field.

The translations are credited to Bill DeSmedt as part of the "Vurdalak Conjecture" website.

Here is an alternative look at the Tunguska event including a collective twenty four witness accounts set down in three parts, the Russians, the Evenki and the Westerners, the first two being actual observers of the event. There is a great deal of data available online for "Tunguska" (of mixed sanity), but thoroughly reading through the witness accounts listed there is worthwhile. The website is well enough informed but tends to an entertainment value. The term "soapbox seminar" is directly used. The purist scientific papers are worth seeking out but do not make light reading either for technical content or uneven translations from Russian.

Sergey Oldenburg, Secretary for the Imperial St Petersburg Academy of Sciences made an official enquiry to the Governor of Yenisei Province, an individual named A.N. Girs. He was probably the highest government official situated closest to the event. The St Petersburg press had published an article on the events. The Governor was already in possession of a report filed by a police officer named Solonina concerning a "bolide" seen above the town of Kezhma that morning. ("Bolide" and "aerolite" are generous translations.)

The report described the great height and size of the body plus loud accompanying sounds like gunshots. Girs responded that he had no information but ordered the police at Kansk under an officer named Badurovto to look into the matter. He in turn could not substantiate the story. By now it was October 1908. In evaluation, it would seem that here was another individual, this time in a position of formal political power rather than academic influence, who preferred a quiet life. There was no further local official enquiry into the matter. In 1908 not even the St. Petersburg Academy of Sciences was disposed to any further action or studies concerning the fall of the meteorite.

The St Petersburg Academy of Sciences (which became the Academy of Sciences of the USSR) was recognized as the "highest all-union scientific institution." Oldenburg himself met with Lenin in 1917 to discuss the future in the uppermost levels of political organization and economic resources allowed it by the new Soviet regime. It was moved that the new Academy would report to the Department of the Mobilization of Scientific Forces of the People's Commissariat of Enlightening, replacing the Provisional Government's former Ministry of Education. By 1929, an official commission was busily purging the counter revolutionaries in its ranks. Oldenburg was one of those relieved of his position in the Academy. In 1934 it moved to Moscow, where, among other scientists, Kulik was later employed. It is recorded that his study was also his accommodation there. A lifetime later, in 1991, the institution reverted to the title of the Russian Academy of Sciences by presidential decree.

A natural H-bomb occurring in the sky is a peripheral suggestion for an explanation. Whether such a thing can occur outside stellar interiors or terrestrially at all is a broad question. It certainly is not common if at all possible and has been termed the "atomic heresy" in our context. The subject of nuclear explosion is generally not something governments want the average citizen to know too much about, anyway. How this strange bolt from the cosmic blue produced such explosive forces and high temperatures remains the real enigma. What this internal source of energy comprised of is a pivotal question that we will address. Chemical, nuclear or some other form like electrolysis or a vapor explosive release are alternatives for the source of energy.

The Barringer crater in Arizona, probably the best preserved of any such large natural scar on the face of the Earth was formed by impact with a major object about 50,000 years ago in a 3.5-megaton event. Its dimensions are ¾ mile wide and 570 feet deep and it is a much visited tourist site. An estimated 300 million tons or rock were thrown into upheaval as it struck. It is relatively pristine as these features go and clearly was formed in the very recent geological past. It was formerly known as the Canyon Diablo and was ascribed a volcanic origin, a good early guess as it lies only 40 miles west of the San Francisco volcanic field. The original meteor clearly made it from space to the ground intact, vaporizing almost entirely as it hit. The simple situation of a single giant crater and a few meteorites left over tells us that.

Kulik was aware of Moulton's studies there and knew that American researchers had established the case for impact rather than volcanic origin as the cause of the Barringer crater. The conclusion that there was no giant iron meteor buried there was much to the cost and disappointment of Daniel Barringer himself. He had invested on the hope of recovering a major find. The speed of impact has been revised downward to some 28,600 mph from a previously held value of 45,000 mph.

In July 1994, the Comet Shoemaker-Levy 9 entered the atmosphere of the planet Jupiter and spectacularly broke up. The views from the Hubble Space telescope of the comet fragmenting into 21 pieces in the huge upper gas clouds of Jupiter gave astronomers a unique show. Each fragment seems to have had its own nucleus as it took its terminal dive into the colossal atmospheric mantle that covers a planet 80,000-miles in diameter. There were impact disturbances to the great upper reaches larger than Earth. It was a very spectacular. Another object suddenly impacting the south pole of Jupiter fifteen years later, in 2009, was likely an individual asteroid, as indicated by the lack of any comet-like trail. Spotting that was a great and fortuitous triumph of equipped amateur astronomy and congratulations go to Anthony Wesley and his homemade 14½ inch reflector telescope in Australia.

The butterfly shape of the forest blast pattern in Siberia may be cited as evidence of an actual nuclear explosion — a major inconsistency for the year 1908 in itself. The epicenter lies near the head of the two lobes. Nuclear weapons were not developed until the Second World War, of course, or witnessed until 1945. In the order of technological progress any atomic theory for Tunguska is post war in origin. The "meteorite" label was originally given in the newspapers of Siberia and Kulik clung to the explanation. There was no other explanation proposed for half a century. We will also hear of genetically-altered plant and insect life, an accelerated growth of surviving trees and vegetation in the region and reports by one source of higher background radioactivity measured decades later.

Alexander Kazantsev (1906–2002) was a high level engineer and colonel in the Soviet army and as a creative writer had won a film award in 1936. He visited Hiroshima as member of the Soviet team. He certainly grasped that the trees at the epicenter in the taiga had been standing perpendicular to the front of the blast wave. This explained the slightly lesser destruction at the center of the blast zones. The similarities were quite apparent with the two explosions. He advocated that the Tunguska blast similarly took place miles above the ground. It was as a literary device rather than an academic paper that he first suggested a spaceship origin to the Tungus event. The Soviet academic establishment (such as the Committee on Meteorites for the Academy of Sciences or KMET) disparaged this line of thinking. The spaceship hypothesis never met with official approval, but it is so appealing that it cannot be stamped out merely by dedication to the scientific mindset.

The Independent Tunguska Exploration Group was formed under Gennady Plekhanov and Nikolay Vasilyev in 1958 and its small group was given the responsibility, among other objectives, of settling the crashed spaceship idea. Dr Victor Zhuravlev was another principal investigator. There was already a kind of "invisible college" for the subject, wholly independent studies and discussion groups involving some competent intellectuals and teachers. The idea had a peculiar popularity in the USSR at the time and still has a kind of cult interest.

Plekhanov's party of summer 1959 duly arrived for a scientific field excursion but the members were initially involved in battling forest fires. When they were able to commence scientific studies, they established after a month's equipped research that the level of radioactivity at the middle of the Great Hollow was larger by a factor of two than that at its periphery. The analysis of the soil samples and tree ashes revealed trace higher concentrations of rare earth elements, facts which enliven the "atomic heresy" idea. Or there could have been a brief hole in the ozone layer above the immediate environs, temporarily allowing a greater measure of radiation to reach ground level from space. No one has mentioned, though, a 50-mile high black cloud that would have accompanied a Tunguska ground zero.

Speculation of the intensity and duration of the temperatures involved is a standing problem. It was certainly more intense than a regular forest fire. Magnetite and silicate spheres, possibly of cosmic origin were located but may not be assumed as fragments of the TCB, being indistinguishable from the ubiquitous meteoric dust. From a eleven page 1998 paper by Vasilyev, one of the best hard science evaluations available, he commences section VI "On the substance of the Tunguska object" with the statement that the diligent search for large fragments of the Tunguska cosmic object which had begun in the late 1920s and ended in 1962 has shown totally negative results. There have been no traces of astroblemes, an eroded remnant of a large impact crater. Another tract tells us:

Increased concentration of microspherules enriched with copper, zinc, gold and some other volatile and chalcophile elements is found in the 1908 layers of peat and wood resin at a number of locations in the region (Longo, G et al., 1994; Kolesnikov, E.M. et al., 1977). Cosmogenous nature of these anomalies is probable but they should be differentiated from aerosols produced by the burning of peat (Doroshin, I.K., 1988) and (possibly) wood as well as from volcanic ash.

It is also asserted that the finely dispersed silicate and magnetite space material present in the soils and peats in and around the region of the explosion do not directly link to the TCB. There is good reason to believe that it was the variations in the background fall of space dust. The radioactivity at the epicenter is within the range of fluctuations of the present background radiation but simultaneously the magnitude is somewhat higher at the epicenter than at the periphery of the region. Academician Vasilyev concludes that most radionuclides concentrated in the upper horizons of the soil and peats were accumulated by global fallout following nuclear tests. Finally, it is noted that the isotopic composition of inert gases accumulated in rocks near the epicenter does not reveal any peculiarities that could be explained by the action of neutron irradiation on the natural environment at the epicenter.

ITEG was successfully funded for a second expedition on the strength of some fresh results. The subsequent expedition had better trained technicians and concluded that radioactivity levels were greater than normal though only marginally higher than the normal background level. Fallout from Soviet nuclear tests was possibly a culpable cause. The researcher Dr Alexey Zolotov (1926 -1995) who has been noted to be as enigmatic as the subject matter we pursue was convinced of a nuclear catastrophe taking place and went considerably further with his speculations. He had a forcefully individualistic approach and incurred more than a little opposition. In his paper "The Problem Of the Tunguska Catastrophe" 1969 the spaceship theory was strongly propagated.

We can put the question simply enough: was the explosion chemical or nuclear in type? It was inexplicably forceful for a large exploding meteor. K.P. Florenskiy's major report published in *Meteoritica* Vol XXIII (1963) holds that a nuclear origin is a "fantastic" suggestion based on "questionable competence." The text generally and specifically criticizes Zolotov's group, e.g., he "submitted hasty and unfounded conclusions on the basis of extremely inadequate and random data." Florenskiy does concede that there is some advantage to sensational publicity but warns that it should not be a basis for the furtherance of scientific knowledge.

Thermoluminescence dating is the determination by means of measuring the accumulated radiation dose of the time elapsed since a material containing crystalline minerals was either heated or exposed to radiation. In usual studies this is regular sunlight. It is a technique used in both geology and archeology when carbon dating is not applicable. Very weak radiation traces do exist at Tunguska ground zero and the TL at the epicenter is indeed greater than the random background level. It also lies along an axis of symmetry consistent with Fast's second estimate of the TCB's trajectory that we will shortly describe. Vasilyev remains puzzled why this effect manifested itself only in the areas of the light flash and not in the entire zone of the forest fire. Incidentally, I've found mixed reports about the duration of the original fires in 1908 and accept a consensus of four days. Boris Bidyukov made headway in discovering traces of the hard radiation from the explosion and founded the "Tungussky Vestnik" journal. Laboratory work on rock and soil samples give supporting evidence. The zone of the radiation burn is much smaller than the leveled forest.

In 1963 Wilhelm Fast, a mathematics prodigy from Tomsk University addressed the matter of the pattern of fallen trees far more methodically having become interested in the investigation of Tunguska.

Dividing the grid into over 1000 test areas each 50 X 50 meters in size and containing 100-400 trees each, the azimuths and angles of the lying trees could be quantified. 650 test areas and 60,000 trees were measured. Based on the measurements painstakingly forged ahead every summer for 20 years, a spread-eagled butterfly shape duly emerged. It bore a wingspan of 44 miles and a body length of 34 miles and was referred to as "Fast's butterfly' having at its center an "epifast." The entire zone covered 820 square miles. Fast has even been called the Newton of Tunguska for his remarkable mathematical work that finally became the first formal scientific dissertation on the subject. A detailed map as part of the process was published in a 1964 article in *Meteoritika*, the publication of KMET. No such elliptical shape was revealed that an exploding regular meteorite would have wrought on the forest.

Fast had first attended a meeting on Tunguska by accident when ITEG 2 was in the planning stage. He had assisted with translation work and when he obtained university support to produce an academic dissertation he took a strictly mathematical approach to the problem.

The axis of symmetry revealed lies at an azimuth of 115º suggesting a flight path ESE to WNW in direction, although such conclusions were excluded from Fast's supervision of the project. In moving from the epicenter towards the border of the leveled forest area, the more consistent becomes the radiality of the fallen trees. Further away as the blast wave weakened, the trees are felled in a more chaotic state. The axis of symmetry was later revised by further study to an almost E to W trajectory. This may be a correction or it could further open up the possibility of a second body in flight. It could have split in two. Fast modestly put a zero emphasis on developing any physical models.

By the mid sixties, another ITEG member, John Anfinogenov studied the central area of the destruction of the forest. They were equipped with the results of aerial photography conducted as part of a more general State program. Their map closely agreed with the previous unassailable research. This second descriptive term "Anfinogenov's butterfly" emerged, also bearing an axis of symmetry from practically E to W in direction but a little different to "Fast's butterfly." A strip of living trees mixed with the "telegraphnik" runs west from the epicenter. Inexplicably, some are almost untouched. Subsequent work by Fast's team concurred in turn with an east to west symmetry indicated in "Anfinogenov's butterfly." This does not simply correct the initial ESE to WNW assumed direction of flight. It could indicate the trajectories of two bodies. The work is crucial to any understanding of the Tungus mystery because much may be deduced for the position and final path(s) of the TCB as it erupted above the forest.

A weak herringbone pattern running for 12 miles in the western part of the forest. This feature could be accounted by part of the surviving TCB travelling on in this direction. There is a slight deviation of fallen tree patterning from radiality although Zolotov's estimates support Fast's initial trajectory. It also presents us with a lively discussion - whether the body had a low angle of approach with a blast wave in excess of the ballistic shock wave or a high and steeper incoming path where the ballistic shock wave exceeded the scale of the blast wave. Overall, a low speed and low angle of approach are favored with greater impact wrought by the blast waves. That wave took an estimated 10 seconds to reach the ground. Taking the Zolotov theories at face value, we have a low angle TCB and the scale of the blast wave exceeds the magnitude of the ballistic wave. We also derive a low velocity body, in the order of ½ mile a second terminal speed. In turn, there can be very little kinetic component to the body's explosive force so the bang was delivered from its internal structure.

Both presumably impacted the forest. Which did the greater leveling damage and in what exact order of swiftness remains a discussion. Clearly the hot blast wave caused the burning but did it partly or entirely flatten the trees? By Fast's invaluable work the whole and completely detailed patterns of the trees' destruction has been preserved before it became too late. Growth proceeds in the taiga, as ever. Throughout the research, not a single meteor fragment is ever found, of course. Here the mystery dramatically deepens. Are there two TCBs or more to ultimately contend with at ground zero, an A and B body?

That the flash came from an estimated 5 miles altitude is again indicated by the work by Zenkin and Ilyin of ITEG. It was a very uneven phenomenon too. The intensity of the thermal burn was far more irregular than originally supposed both on the trees and ground moss. This cannot be explained merely by the local topography or remotely by the weather conditions at the moment of irradiation from above. The evidence suggests a very non-uniform pulse of energy of uncertain wavelength.

To the early explorers, the burn effects were noticeable on not merely trees and bushes but on the marsh moss. Larch trees are comparatively sturdy in resistance to fire and survived the better. Birch, aspen, alder and dark conifers are more easily damaged by heat. Kulik had noticed the extensive burns covering all the vegetation including the marsh moss but had no concept of a great "light flash" but rather a simple fireball. The zone of the light flash is considerably less than that of the leveled forest. It measures approximately 11 x 2 miles with an axis of symmetry running E to W. The two possible trajectories illustrated by Fast's work are borne out again. The center of explosion does not coincide with the center of the flash, instead it lies about a 1¼ of a mile from the "epifast." Its egg shaped contour is flatter to the E and more pointed to the W. Studying the comparative intensities of the energy flash produces a far more complex result. Paradoxically, the burning was weaker under the trajectory of the TCB than at distances from it. These studies on the burn phenomenon do conclusively show that the path of the TCB terminated over Mt. Stoykovich and that the forest fire started simultaneously. The pattern of the fire was quite irregular due to the shape of the terrain. It hardly extended beyond the zone of the leveled trees and was about five times the extent of the light burn. Further calculations indicate that the light flash comprised a mere 10% of the total energy expended in the explosion.

This is probably more energetic than any chemical explosion can provide. Two bands of trees not lying strictly radially are known. This could indicate that one piece of the TCB ran from the ESE and another continued from the east to the west. Fast's analytical work stands up splendidly and the TCB splitting rather than erupting in an homogenous blast are valid conclusions. The zone of the radiation burn lies almost directly east to west with the dimensions of slightly more than 7 by 11 miles and an area of 80 square miles. The proportion of radiation emitted as light may be as low as 10–25%.

The forest fire itself was limited in area and probably did not even reach the boundary of the leveled trees. Burning branches and debris were scattered some 20 miles from the epicenter and in all directions. Both the fireball itself and intense forest fire joined to create a huge pillar of hot air. This presumably changed the direction of the local winds that began moving to the center and restricting the spread of the firestorm. From the observers at Vanavara we accept a dark cloud 50 miles in height. At greater distances treetop and some ground fires persisted for a collective five days, another source informs us.

We have naively assumed that there was one body in a single explosion and no maneuverings of approach. This could be an oversimplification. Going all the way back to the eye witnesses, multiple bangs were heard, never ascribed to echoes although a single sonic boom could be theorized. A shifting trajectory might have been the case and remains one of the insoluble problems. One disappointing consideration on the sightings of 30 June 1908 is that the accounts of the path taken throw up contradictions when closely studied together. There is no proposal of multiple bodies appearing in the first place and we deem that the accounts are sometimes confused and inconsistent. One really can only go so far with them. We always favor a lower speed because a brief time did allow people to see something coming, unlike a fast action TCB that would have been here with minimal visible warning.

The placement and irregularity of the burns therefore suggest two bodies at the end. It could discourage a nuclear pulse for a likely hypothesis because they are more regular in their thermal effects. The atomic fireball propagates as a sphere. Using the computer tomography method and applying the shape of the ground burns there was an attempt to investigate the real shape of the light source. Ryzin, another ITEG member, here obtained a truly remarkable result. Instead of ball, sphere or cylinder the source of the light flash looked like the top of a mushroom with a convex surface at the top and a concave one at the bottom. We would dearly like to know the wavelengths associated.

Fast did not use the term "epicenter" as he regarded even that as a mathematical abstraction on his meticulous survey. He is admiringly quoted, directly from Newton, stating, "I am not interested in hypotheses!"

In the order of events, the Soviet space effort was going from achievement to more spectacular achievement in this era of the late 50s to early 60s. The Russians flaunted their technical superiority over the West in the form of the *sputniks*, Moon probes and the first manned spaceflight by Yuri Gagarin in April 1961. The very first probe, Lunik 1, launched in January 1959, passed within 4,000 miles of the lunar surface and moved into permanent orbit around the Sun. It is still out there, somewhere. It was more for political/military posturing than scientific research that Kennedy inspired the United States to "go to the Moon" in his evocative speeches. He was looking for a high technology race with the Russians that could be won. It took a long time to catch up, and he had made the challenge clear. It did not come out until years later that the Russians had major problems with their N2 booster vehicle that in four separate launches never achieved a test flight. In 1968, two such major booster failures resulted in explosions, scuttling any plan for a crew to join the orbiting vehicle for a Moon mission. Their equivalent lunar descent module had been tested in Earth orbit.

The American Skylab was placed in orbit during 1973 and was visited by crews on three occasions. It remained aloft until 1979. The joint Apollo–Soyuz Test Project single flight in July 1975 was a bright illustration of détente. As planned, it was America's last manned flight until the first space shuttle mission of April 1981. The Russian space station Mir, commenced in 1986, retains the record for the longest continuous human presence in space. Its rotation of Russian and American crews ran Mir for eight days short of a decade. It was deliberately deorbited in March 2001.

The Russians have never been beyond Earth orbit. No one has since Apollo 17, that last Moonshot of December 1972, for that matter. A little-known fact from the Moon missions was that the Russians with their unmanned Luna 15 of July 1969 were literally mere days ahead of the Apollo 11 mission. They must have been trying to upstage the American space effort in successfully returning lunar rock samples before the one giant leap for mankind that included some geological prospecting. Apollo 11 duly returned the first moon rocks to Earth. Luna 15 crashed on the lunar surface and was entirely disabled. The Russians subsequently had three successful sample and returns from the Moon's surface. Needless to say, these scientific efforts took precedence over research into Tunguska.

The early expeditions to Tunguska had noted the rapid growth of the young trees then 20 years on. Heat sterilization of soil can explain the subsequent faster growth, a condition noted in the surviving trees following more conventional forest fires. Apart from ash fertilization there was both a decreased competition for sunlight and a greater availability of ground minerals following the event. "The boundaries of accelerated forest growth follow the boundaries of the fire and flattening of the forest quite closely" goes Florenskiy's major report from 1961 continuing to "express grave doubt as to the possible participation of meteoric matter." It is found that the accelerated tree growth is a feature of the ecological conditions. A specific conclusion that there was no change in the growth rate of bog communities beyond the moss that grew directly on the mineral-rich 1908 layer over which the fire passed.

One discovery from 1958 reveals that at the central part of the forest both surviving and newly grown trees there were unusually wide growth rings up to 9 mm as opposed to mere 0.2–1.0 mm. The novel term "cometary fertilizer" has even been coined but it strictly lacks direct chemical evidence. They are mostly Siberian spruces (Picea obovata), Siberian pines (Pinus cembra), birches, aspens and cedars in the region. Today, the forest has made a pretty full recovery but visible tree damage on the ground and charred pieces from that violent dawn may still be found underfoot. The remains of shattered tree remnants still lie in a direction oriented away from the epicenter. The conifers are most prominent after all this time and most of the measurements applied to them. The fallen broadleaf trees, generally the aspens and birches, are mostly rotted away now.

On the subject of genetic disturbance, pines with 3 instead of the usual 2 needles in a cluster do occur more often in the Southern Swamp, their number dropping off with increased distance from the epicenter. The very greatest numbers of such pines occur coincident to the maximum quantities of trace ytterbium on the Ostraya mountain zone. The second largest concentration of affected trees grows in the canyon where the Churgim Creek flows. It has been shown to be an inherited and genetic trait.

Among the ant population present the researchers Dmitrienko and Fedorova discovered further genetic variations from the norm in both head and eye sizes. In compiling the prodigious Atlas of Genetic geography of the USSR the geneticist Rychkov concluded that a human Rhesus negative individual named Olga Kaplina had inherited her trait from both parents being witnesses to the Tungus event. The trait is rare in mongoloid Siberian people.

Conditions that may be interpreted as radiation sickness in later years were rumored among the local population. Surviving reindeer carried "scabs that never appeared before the fire came." The wounds may also have been simple burns sustained in the burning forest. A Russian scientist named Krymov was actually Tungus born and clearly recalls that his father went into the fallen taiga and a few days later died in terrible pain as if he were on fire. The young man's later science education encouraged him to conclude that it was caused by radioactivity. More colloquially, the god of fire and thunder legendarily burns with an unseen flame within. No wonder the area was shunned as cursed by an evil omen from the sky. There was, unfortunately, no timely study on the diseases or disorders allegedly caused by the Tunguska explosion and no medical records to study. The chaos, war and deprivation that consumed mother Russia may never be fully documented for sheer scale and as we note, the Tungus event could have slipped into complete obscurity. It was hardly a high priority at any time. On 30 June 1908 fears that this was the end of the world had gripped some people in the moment. It was interpreted as a divine punishment for their collective sins and a reaction of religious wrath and retribution from on high. In the village of Korelina a deputation visited the archpriest seeking advice on how best to prepare for it. What they were told was not recorded.

That some Evenki tribesmen hid their knowledge of the great fall and were both fearful and secretive about it was a circumstance. The people could be reluctant to give information or guidance to researchers or outsiders, as was noted by several working researchers in the earlier days. That they were somehow concealing a great meteorite itself cannot be a material fact. We can be confident that any genuine piece or specimen would have been revealed by now and indubitably a giant space rock at rest in the region would have been located. The shaman Vasily was apparently killed by his own people for his involvement in helping to guide our man in Tunguska to the center of the presumed fall zone. Later, one member of the expedition team was found dead under mysterious circumstances. Both Evenki belief and Soviet national secrecy obstructed any in-depth study of Tunguska.

Scientists deplore both ignorance and superstition on one hand and politically motivated suppression of scientific information on the other. That certain data may have been concealed for decades is a possibility. However, much worthwhile work exists that has not been translated into English and there are long lists of research papers devoted to the subject. Some are specifically focused.

Suslov's determined enquiries and fieldwork on the Evenki religious beliefs included the accounts of dark, metal thunderbirds called the Agdy whose eyes flash lightning and wings release thunder from their flights through stormy skies. He was told that years of tribal feuds culminated in a uniquely gifted shaman orchestrating a legion of them to descend on the land of the Shanyagir clan. The ancestors of the thunder live in Khergu, the lower world and habitation of the dead. One Tungus named Onkoul lost 250 reindeer in the disaster. Suslov was once informed when he could get any Evenki to speak of the event. There was a powerful noise and crevices appeared in the ground. Only the Agdy can now live at the place of the catastrophe and no one dares enter the area, we are admonished. A famously powerful shaman named Magankan was chiefly responsible, it is revealed, an alleged miracle worker.

Suslov had a scientific motive with his research, the ambitious hope of locating the meteorite. He received persistent silence and the deflection of questions as his surviving texts, made available in translation in 1983 attest. He played a major role in the establishment of Soviet power in the region, following the Russian Revolution. The material of his 1920s fieldwork are available online, translated by Joachim Otto Habeck. The wizard-like omnipotence of Magankan included the working of miracles and being bulletproof. Also, we read in Suslov's notes that:

The members of the clan Shanyagir ascribe the impact of the famous Tunguska meteorite, which at present is searched for by L.A. Kulik between the Stony Tunguska and the Chunya (rivers) to Mungankan too.

Crediting superhuman control over nature to charismatic individuals forms a common thread in primitive religion. Mythology and history have their gray areas. Science opens up a vista of nature over and above superstitions and divine explanations.

As early as 1931, Kulik had remarked on the thickness of the growth of the bog moss and vegetation among his reports that appears at odds with the findings of the early 1960s. Companions quite familiar with the terrain and natural landscape made similar observations during that early era of investigation. Presumably, the initial recovery was marked but settled down to a more regular growth. The July 1928 issue of *Scientific American* contained the first published study to appear in the Western press. *The Literary Digest* also warned on the dangers of major meteor falls with this as the sobering example. They printed an interview with Kulik and Sytin as part of the coverage given the second expedition of June 1928. The New York and London newspapers also printed columns on their progress.

In the United States both the Mount Wilson and Smithsonian Observatories noted a sizeable decrease in the transparency of the atmosphere for several weeks in July 1908. As noted, they were particularly astute on this as a regular observation and measurement. There were magnetic disturbances on telegraph lines around Prague and an observer there distinguished that the strange twilights were not like an aurora in appearance. Letters to the newspapers commented on the brightness of the nights and noted that one could read large print at midnight or could have continued late with a round of golf. Photography was possible far later in the evening by a strange extended length of daylight that looks eerie to this day. Evening trains were able to run without lights.

Dimmed sunsets were seen as far away as California two weeks later. We also deduce that the air heated by the explosion produced enough water in the atmosphere to engender noctilucent clouds over vast areas of the sky. These clouds are typically 47-53 miles in altitude in summer, occurring between latitudes 50-70 degrees north or south of the equator. They are the very highest mesospheric clouds, normally too high to be seen. They are visible only when illuminated by sunlight from below the horizon while the lower layers of the atmosphere have entered the Earth's shadow. Their source is water collecting on dust particles.

The tiny remnants reached enormously above the stratosphere, reflecting sunlight from above the Earth's shadow line.

A. Schoenrock, Director of the St Petersburg Physical Observatory, had considered three possibilities for the huge illuminations covering about 25% of the horizon in 1908. The aurora borealis or northern lights to which the glow bore little resemblance, high altitude clouds or a massive injection of dust in the upper reaches of the air. He was not at the time aware of the explosion and came to no specific conclusion. The strange and fluorescent particles in the atmosphere were not scattering sunlight like a massive quantity of dust or volcanic ash would appear. It was reported in the St Petersburg newspaper on July 13 with reference to "lucid twilights" quite unlike the aurora.

In the early 1990s, V. Bronshten photometrically analyzed a large sample of contemporary photographs, concluding that the illumination had been greater than a normal night sky by a factor of 100. The higher levels of brightness paradoxically were occurring a long way from their presumably Siberian origin. Romeiko applied even more advanced digital techniques to the pictures shot at the time including astrophotography and derived an even bigger factor of illumination compared to the regular dark sky.

A peculiar fact emerging is that a major dust cloud was passing over Mt Wilson and was first detected on 4 June 1908, returning on 4 August and 4 October, apparently with a 60 day period. Note that the first occurrence was prior to the Siberian fireball by date and quite independent of the differing calendars used in 1908. It was deduced that another major meteorite had plunged undetected into the atmosphere over the Pacific Ocean, completely expending itself without impact but leaving a major cloud of meteoric dust. Mt Wilson Observatory regularly studied the atmospheric transparency at several wavelengths, as did the Smithsonian facility. That dust cloud is not presumed related in origin to the TCB.

Note also that it was on the date 17 June 1908 that the event took place according to the Julian calendar still used in Siberia at that time. In Russian history, the October Revolution happened during November 1917 according to Europe, where the Gregorian calendar had been employed since the later 16th century. The Old Style calendar applies to the event and the dating of newspapers and letters authentic to the period and location. Christmas Day in the Russian Orthodox Church still falls unchanged in January by date, not 25 December. In London it was 0:17 hours GMT and in New York just after 7 pm on the previous day, 29

June at the moment of the explosion. Subsequent studies later refined these precise timings to an uncertainty of a mere 10 seconds. The calculations from Irkutsk had given 0 hours 17 mins 11 secs in Greenwich Mean Time.

Unusual atmospheric phenomena had actually been observed prior to the event, a circumstance adding to the enigma. Lengthy twilights and high altitude silvery clouds were later found to have been seen across an area westward of the fall site with an intensity that seemed to increase from east to west. A huge aurora borealis was not sufficient explanation although given in some instances as the prime cause. Whilst the far northerly extent went unrecorded, none were seen within almost 400 miles of where the fall site would occur. This is a strange dearth of observations in the local territory shortly destined for the impact. Somehow the prior effects were not seen in the very area the main event was to take place.

The ITEG concluded a remarkably detailed piece of work in 1963 entitled "Noctilucent clouds and optical anomalies associated with the Tunguska meteorite fall." The extreme brightness of the night June 30/July 1 was corroborated by a wide range of periodicals, newspapers and ships logs by enormous research work. How the Earth's atmosphere may have selectively and regionally been prepared for the event in advance remains deeply mysterious. We are clearly dealing with something quite extraordinary to any discipline of geophysics, meteorology and astronomy. The phrase "Tunguska precursors" enters our lexicon. Robstov does a fine job in dividing the impact of the TCB into three connected keys, the leveled forest, the light burn and the storms in the magnetosphere that we shall address.

Kasantsev was struck by the "telegraph pole" nature of the trees stripped and burned but left standing at the epicenter of the in the forest. As noted, it was not unlike the partially surviving buildings at the very center of the horror of the Hiroshima explosion. Dynamically, this is explained by the huge but solely downward forces accompanied by comparatively little transverse action at ground zero directly beneath the inferno. A more horizontal wave is the more thoroughly destructive.

Simulations and all studies agree that the TCB gave rise to a powerful airburst at about five miles altitude. The blast took place aloft and this is an uncontested fact. Both thermal and ballistic waves are clearly involved to have scorched and flattened millions of trees in a very brief time. The exact sequence is still debated but by clear evidence, the burning was very swiftly put out, clearly by another wave. As we shall see, the speed and angle of approach are open-ended questions with a consensus forming around lower values for both. The generally accepted conclusion is a few miles a second or less, for speed, with an approach angle to the ground of no more than 30º. Exactly how brief and intense and from what source came that burst of energy remain the formative questions. A distinct explosion occurred or possibly a series in quick succession at the final moments. It was not a case of gradual ablation with the body gradually losing mass until little was left as happens with smaller meteors. The height at which an explosive fragmentation could take place depends on a wide range of factors. The entry angle, speed, size, shape, composition are, of course, among the parameters and only the first two can be roughly assumed from observations.

Beyond the epicenter lay colossal patterns of charred and felled trees up to nearly 40 miles distant. The series of photos from the late 1920s clearly convey the scale of the destruction. Nowhere else has such a huge quantity of toppled trees been recorded. Four radial patterns were formed among the downed trees that presumably correspond to the series of great bangs "like cannon fire" possibly prior to a final grand or series of explosions. This was shown by the early expeditions by some skilful scientific detective work conducted among the damaged timber. It is also consistent with the peaks in pressure recorded by the barographs readings at greater distances. The forest looks like so many overturned matchsticks with a force of destruction in a single moment that no one, prior to Hiroshima, could grasp. It is clear that relating the explosions to cannon fire is simply the loudest bang those Tungus natives could mentally grasp beside strong thunder.

Collectively, there were several cacophonies of sound apart from the acoustic waves of approach that took time to reach the people on the ground. One witness referred to five thunder strikes but strangely, the final one was the least loud to his ears. The number of bangs sustained and their relative intensities as experienced vary from source to source. They may have been louder in the area north of Lake Baikal. There is no way of knowing how many occurred above the Great Hollow. The reports are quite haphazard. One account held as many as fourteen bangs taking place and the visual reference cited a "pillar of fire in the shape of a spear." It is reasonable to deduce that the source of sound was moving from east to north from witnesses located closer to the fall site. Loud noises were heard in the whole area between the Yenesei and Lena Rivers that geographically defined the Tunguska area's western and eastern boundaries, covering about 300,000 square miles. Apart from being compared to thunder and artillery fire there was a loud knocking as if large stones were falling. The TCB was heard over a far greater area than it was seen. The zone was approximately 900 miles wide as estimated, a circumstance easily deduced by the curvature of the Earth, additionally suggesting a low angle of approach to the ground.

What, then, was its speed? All we really know is that it was faster than the speed of sound yet short of a cosmic velocity. There was a visual warning of its approach. At Earth's distance from the Sun and in a stable orbit, this planet is actually moving at 18 miles a second. The body may have caught up with the Earth allowing a reduced speed of interception. Equally possible, the TCB ran head on into us at comparatively high velocity. A speed of approach as high as 40 miles a second or more could be hypothesized in that case and if so, the object greatly decelerated on approach. There may have been some changing of course during the descent and quite possibly some major veering during its final miles. This is consistent with normal meteor studies and explosive fireballs. Of course, the precise direction from which it originally came cannot be known. We can only reliably say that it arrived from the SE on its final leg. The ITEG group concluded a TCB visible for a five-minute period based on the collective accounts. It was not shining brightly until the last stages of its display. Speeds of approach for meteors angling up to or running head into the atmosphere are well understood by radar observation.

Normal incoming meteors generally begin to emit light at heights of 80 miles or less. This renders a low average speed of only a few miles a second as atmospheric drag takes effect, far slower than the cosmic velocity typical of an object in orbit around the Sun at our distance. A more elongated trail of damage to the forest would have resulted if the final cataclysm had been from a higher speed body, we reason. If it were first seen some 400 miles from the fall point (a vague assumption equally based on the timeline of the body's appearance), then we could conclude a low velocity and shallow trajectory from the short time taken to cover the distance. It came into view at some unspecified point north of Lake Baikal, by report. From this initial sighting at an unknown altitude it headed at unspecified speed in the direction of the Stony Tunguska River. Eye-witness reports are all too hazy and there are no reliable sets of hard figures to work with. It must have been visible earlier than those surviving and recorded sightings convey but such data is not available. We have no solid values like the right ascension and declination positions, visual magnitude and angular shift from its initial appearance.

If we knew more about the velocity and trajectory, we might be able to eliminate any hypotheses based on a high speed, high angle model and suggest that the force of explosion was integral to the body with very little kinetic component. This is an appealing line of reasoning but raises difficult questions regarding the mineralogical and chemical nature of the projectile, leaving us with a deeper conundrum. A simple meteor or cometary core does not fit the facts of such a powerful explosion. Absolutely no remnant pieces shearing off and blanketing ground zero or the path of approach were ever located. We must reluctantly conclude that they do not exist. The TCB survived intact until the very last moments with no trail of debris or craters forming as it drew nearer. It has been called traceless Tunguska. There is no evidence of a process of gradual ablation, splitting up or losing mass as it descended, which any such object is calculated to undergo at least to some extent. Therefore a small object of high density and great tensile strength is suggested, moving comparatively slowly.

We could assume it was a natural space body of asteroidal or cometary structure with an irregular shape and some element of spin. It was obviously brightest and at its most imposing along the final track. Towards the end, the inhabitants of Vanavara directly felt the passing heat. It survived until it reached about the height of Mt. Everest, entering from space at what probably finalized as a low angle of approach. It held itself together as a single entity until the very end.

June 30 had been a clear dawn with the exception of a small dark cloud reportedly lying in the direction from which the fiery body came. The huge pillar of smoke resulting from the final immolation was visible for some 250 miles and was perhaps 50 miles tall. The possibly mushroom-shaped great cloud was reaching back up into space itself. The *Krasnoyaretz* newspaper on 13 July 1908 records that:

> The sky at the first sight was clear. There was no wind and no clouds. However on closer inspection to the North, i.e. where most of the thumps were heard, a kind of ashen cloud was seen to the horizon which kept getting smaller and more transparent and possibly by around 2-3 pm completely disappeared.

We should not over analyze the meteorology here, but what information does it convey? Eyewitnesses around Kirensk may have seen a dark cloud following the projectile. The object itself reliably passed 80 miles west of the town. The consensus that it passed directly above the village of Preobrazhenka is also good.

The *Siberian Life* newspaper gave a brief report that month:

When the meteorite fell strong tremors in the ground were observed and near the Lovat village of the Kansk uezd two strong explosions were heard, as if from large caliber artillery. [An uezd or uyezd was an administrative subdivision in a previous Russian system of government.]

Most textbooks and references favor a cometary hypothesis for the "Great Siberian Meteorite," "Tungus Event" or plain "Tunguska" as named after the region.

The comet theory is generally given as proposed by the Englishman Francis Whipple in the 1930s; then it was made independently by Astapovich based on the suggestions of Vladimir Vernadsky. Harlow Shapley (1885–1972) should probably be accorded the original credit with a brief mention of Tunguska and its possible cause in a major work of his published in 1930. This is not generally cited in Tunguska studies. Kulik himself, in conversation, discussed that the Tunguska meteorite might be connected with the Comet Pons-Winnecke or indeed may have been the comet itself. A cometary explanation was studied by the American astronomer William Christie of Mount Wilson Observatory in 1942 and expanded on by F.V. Fesenkov of the Soviet Academy of Sciences in the early 1960s.

V. Bronshten of the same august scientific body expounded that the Earth's gravitational field and atmospheric braking of the particles allowed the larger dust particles to shift westward. It took about six hours for the undefined high-flying material to reach the British Isles and Ireland to the far west of the Eurasian landmass. The smaller particulates made the same distances but were less effective in scattering sunlight. A connection to the Comet Pons-Winnecke may be ruled out. Discovered in 1819 and rediscovered in 1858, with an orbital period of 6.37 years, it passed at 0.42 AU of Earth in December 1906 and was receding from Earth in 1908. The June Bootids meteor stream derives from that cometary parent body and there had been displays in 1916, 1920 and 1927. Archivally, they were even called the Pons-Winneckids.

In June 1927 Comet Pons-Winnecke was 0.04 AU distant from Earth, equal to about 14 lunar distances or 3½ million miles and was within the view of the naked eye. This comet is equally as unlikely to have fragmented and impacted the Earth in 1908 as Comet Encke. The orbital studies clearly confirm this.

If you set up a NASA orbit diagram on a computer for a choice of paces and run it back and forth over time, there is no connection or crucial point. Comet Encke's orbit is stable allowing it to be regularly seen and predicted. It is not constructive to suggest that it significantly shifted over the last century.

In 1966, the first English translation of Krinov's book *Giant Meteorites* was made available. As an original researcher from the Mineralogical Museum, his work contains the best record of the early researches both in the field and academically. Acting as Kulik's Number Two at times, he was a leading authority on meteorites. In the grueling subzero temperatures of one expedition he was hospitalized and lost a toe to frostbite. The available footage contains his direct speaking on the subject in later years.

The early researchers Voznesensky and Kulik also used the terms "Khatanga" or "Filimonovo" meteorites respectively. The former is a local name for the upper reaches of the Podkamennaya Tunguska River and the other a regional town. There were several bystanders among the passengers at the Filimonovo railway junction who heard the cacophonies that fateful morning. The train had halted but it is not entirely clear if the flight of the TCB was directly seen from on board. I've not found any passenger witness statements and there are conflicting accounts on that detail. The driver is quoted to have thought that it was a "phenomenon of nature."

In all his writings, lectures, correspondence and published work, Kulik remained convinced that a major meteor had fallen. To him, meteorites and craters had to be in evidence somewhere and would not be difficult to eventually locate out in the field.

Kulik's diary for 13 April 1927 records that "the results of even a cursory examination exceed all tales of the eyewitnesses and my wildest expectations. The devastation can only have been caused by an air wave of tremendous power."

Here is a lengthy extract from a letter dated 23 June 1908 written by G.K. Kulesh of the Kirensk meteorological Station:

> On 17 June (OS) to the NW of Kirensk a phenomenon was observed which lasted approximately from 7:15 to 8 AM. I did not get to observe since after recording the readings of the meteorological instruments, I had sat down to work. I heard hollow sounds but took them for salvos of weapons fire on the military field beyond the Kirenga River. Having finished work, I glanced at the barograph and to my surprise I noticed line after line recorded at 7 AM. It surprised me because throughout my work I did not get up from my place, the whole family slept and no one entered the room.
>
> Here is what happened. At 7:15 AM there appeared in the northwest a fiery column with a diameter of about four sazhen in the form of a spear.

When the column disappeared, there were heard five strong abrupt bangs like from a cannon, following distinctly one after another, then there appeared in that place a dense cloud. After about 15 minutes the same sort of bangs were heard again and after another 15 minutes it repeated as well. The ferryman, a former soldier and in general an experienced and knowledgeable person counted 14 bangs. In keeping with the duties of his job he was on the shore and observed the whole phenomenon from beginning to end. The fiery column was visible to many but the bangs were heard by an even larger number of people. The peasants from the village nearest the city drive into the city and ask: "What was that? Doesn't it mean war?" In the city there were also peasants from the village of Karelain lying 20 versts from Kirensk on the nearer Tunguska river, they pass on that there had been a strong shaking of the ground such that the window glass was broken in the houses.

Of course, anything has a linear rather than angular size when seen in the sky by an untrained observer. One could be disappointed to hear that from a scientist, something like "subtending ½ degree" would be more useful. He goes on to say that a lake may have formed seven versts from the village of Korelina and in the mountains but that the story had not been verified. It may have been a shift to a wetter nature of regular marshland more seasonally occurring in the annual thaw rather than a meteor impact. It may not be factually based at all. Here, a scientifically-educated individual missed making visual contact simply because he was indoors with the windows closed at the moment. As for a potential new lake, a major blow to the permafrost could, hypothetically, produce a crater that shortly could be inundated by water.

I do not know of a single technically-oriented eye witness account. There are no methodical and scientific descriptions of the fall or the passage of the object referring to visual magnitude, angular size or shift across the sky. The continuing text from Kulesh that considers the input derived from the questionnaires:

> The most powerful bangs were the last ones and the vibrations of the air were strong. There were 14 bangs in all and they occurred in 3 stages, the crashes being heard over localities far removed from one another.

And in summary from the same notes, the tale about a lake being formed in the event is held to be untrue. Many noted the resulting dark cloud and it took as long as three days (according to another report) to put out the fire.

Several times it turns up on overheated websites that the Tunguska crater has been "found at last" and that Lake Cheko has been "proven" to form in the Tunguska event.

Chapter 4. The Cometary Hypothesis and the Taurid Meteors

The Taurid Complex is a known series of related meteor streams in planetary space. These streams are best described as long rivers of material generally from the same source body circling the Sun as minor members of its orbital entourage. One of that group, the Beta Taurids encounters the Earth's path reliably from June 5 to July 18 every orbit of the Sun and hence every year at that time. It is an important condition to one version of a cometary hypothesis involving a specific comet. The Taurid radiant was first discovered in 1947 by radar techniques.

Identifying the TCB as a piece of Comet 2P/Encke was first proposed by the Slovak astronomer L'ubor Kresak in 1978. The immediate difficulty with this theory was that there were no accompanying meteors. That we were passing through the annual high point of a stream is a distinct possibility yet there were no observed companions, audible bolides or scattered meteors seen entering the atmosphere or coming to ground at the same time anywhere at all.

The Beta Taurids are a part of the four tier Taurid Complex peaking on June 30 every year and the date is clearly important to the discussion. It could be the single most crucial piece of information in the whole study. Not that we have received major meteorites on that recurring date, before or since. It remains paradoxical to proclaim it part of a recognized group of a daylight meteor shower and yet observe it as an individual. Complete isolation is the case. No other meteor-related events elsewhere in the world were reported for those specific dates.

A presumedly genuine Beta Taurid exploded on a far later date, taking the form of a bolide over New Zealand in early July 1999. A meteorite, crater and all, came to ground in Tashauz, Turkmenistan on 20 June 1998 and the areas of Nash-ville, Tennessee and Portales, New Mexico had been hit seven days earlier by falls. We may assume that they were Beta Taurid meteors, based on the dates associ-ated, rather than sporadic meteors. They remain a lesser-known group with no record of spectacular displays.

We are better equipped to identify their direction of origin nowadays if we apply ourselves to this low-key area of observational astronomy employing radar techniques. There are several bolides a year successfully (if inadvertently) detect-ed by satellite observation. The equipment was designed and launched as mili-tary satellites watching our neighbors. There are, of course, far busier and more prominent meteor displays reliably worth sitting out for in the garden at night.

In a separate case there was the fall in the Nubian Desert, Sudan, in March 2008 that lacked actual witnesses but was smartly recorded by satellite observa-tion as an explosion at 37 miles altitude. Following the event we think no one directly saw, 280 fragments of asteroid 2008 TC3 were successfully recovered in a community search effort. The smoke trail of its passage, slowly distorting in the upper wind was photographed from the ground. This is a case of observing the fall by instrumentation, locating some meteorite fragments in a successful labor-intensive search shortly afterward, then identifying the parent body. There is an image as a tiny point of light when it was still in space and heading toward Earth. The human experience of major impacts, however, is limited and the dynamics of major incoming bodies are theoretical, however confidently asserted.

The Beta Taurid radiant position currently lies at an estimated Right Ascen-sion 5 hrs 18 minutes, Declination +21 degrees 12 minutes in Taurus, subject to precession. (This is the equivalent of latitude and longitude projected on the ce-lestial sphere and integral to positional astronomy and navigation.) Currently, the Earth's orbit merely touches the periphery of the Taurid stream, yet it is notably intense by standards. Meteor activity has been observed from an area over 10° across. Earth was not in any sense exposed to its core during the early 20th century and probably has not been for several millennia past. There is some speculation that the observation of the wondrous celestial effect of meteors pro-vided any of the motivation to build Stonehenge, apart from the prediction and celebration of eclipses.

The shower has two separate branches, the Northern and Southern Taurids, referring to a position N or S of the ecliptic and also known as the Zeta Perseids and Beta Taurids. Their full dates of activity are an observed 20 May–5 July with a maximum occurring on 13 June for the former.

The Beta Taurids are active 5 June–18 July with a relatively flat maximum now centered on 29 June. The nodes and argument of perihelion are continuously shifting for their celestial mechanics. They are both daylight showers. The more concentrated southern branch may continue as the Southern Arietids into one moving radiant. A 7:2 mean motion resonance with Jupiter has been estimated and the swarm orbital period is close to 61/18 years. After the Earth has orbited the Sun 61 times and the Taurids have completed 18 revolutions the same relative configuration repeats. The Taurid Complex is considered to be a very old and established shower and very large as such streams go.

The Sun on 30 June occupies the approximate position RA 6 hrs 40 mins and Dec +23 degrees 8 minutes every year in this epoch. It is nine days after the Summer Solstice at that position. The Sun was therefore about 17º east of the Beta Taurid Radiant at the time and moving away along its daily ecliptic path as the Earth continues its eternal orbit. It can be inferred that the TCB appeared out of this position in the sky but there is no supporting input that it emerged from any specific point.

Concerning practical angular measure in hands-on and positional astronomy, the discs of the Sun and Moon cover half a degree of the sky but create the impression to the eye and human brain of much more by their brightness. From the horizon to the zenith is obviously 90º and a hand at arm's length covers very approximately 10º of the celestial sphere. In the circumpolar constellation Ursa Major, the leading edge of the Plough section are the two stars Dubhe and Merak who are exactly 5º from each other for handy celestial reference. They are charmingly called "The Pointers" because a simple line drawn northward through the pair arrives at Polaris, the Pole Star. You can point one arm at Polaris and another at Orion's Belt, which lies conspicuously on the celestial equator, the great line of the Earth's equator projected onto the sky. Also, the altitude of the Pole Star in degrees is equivalent to the latitude of the observer in the northern hemisphere. With a basic knowledge of the constellations, one could easily figure out one's approximate latitude if cast away on the proverbial desert island.

This Beta Taurid shower is mostly daylight, peaking after sunrise in late June with a maximum hourly rate of 25 as radar tracked. Some schools of Tunguskan thought say, "There it is, we've solved the problem..."

Dawn and daytime meteors tend to higher speeds, as they are indeed running head on into the moving Earth as viewed by the early morning observer and generally come out of the easterly direction. Those meteors seen catching up with the Earth are generally of lower relative speed. Broadly, this is why the very early morning after the point of midnight can be the most practical time to look for meteors. They can come faster and more imposingly in the dawn sky. This particular Beta Taurid group was first located by the Jodrell Bank radio telescope in 1947. They play no role in the lore or history of meteor storms and sightings. Further work at the Jodrell Bank Observatory in 1950 placed maximum activity on July 2 and their mean daily motion was plotted. It is "rather weak and broad" as a shower as described G.C. Evans' study there. A Radio Meteor Project conducted at Havana, Illinois yielded more data on the stream the following decade. It cannot be seen as particularly important or different to any other meteor group and it is purely because of the dates that they are of interest here. In the 1930s Fred Whipple's study of the two nighttime Taurids showers currently occurring in the October–November annual period anticipated the existence of some daytime Taurids occurring May–July. Considering that 10° area, the mean radiant actually lies closer to the star Zeta Tauri. This was not an error as such and the naming after the star Beta Tauri is applicable.

Later work suggested that the Taurid Complex may be the result of a hierarchical disintegration of parent comet(s) breaking up in stages and over many millennia. Several Apollo asteroids may also be associated. For 30 June 1908, the researcher Kresak was confident from bystander's accounts that its apparent direction of appearance was west of the Sun in Taurus. This fits neatly into the case but it cannot be positively confirmed. Its arrival early in the morning, shortly after 7:00 AM local time and with the rising Sun, may add further credence to a low angle of incidence for the body as it came across the dawn sky.

Strictly observationally, the normally 3° radiant suddenly swells to at least 7° around 2 July. Some of the stream orbit might be in its descending node as it crosses the Earth's orbit in June/July each revolution. C.S. Nillson, using radar equipment at the Adelaide Observatory, is credited with this original suggestion.

If, for the sake of present argument, we identify a major Beta Taurid meteor as the TCB, we apparently encountered one big and single piece of what we know to be a swarm. There certainly is a case to be made for this association but there was no similar material on parallel paths to it or apparently spreading out from a distinct point or radiant. (This is an established pattern for systematic meteor observing and how they can appear under ideal conditions in the night sky.) Obviously, only very major meteorite displays or individuals are visible in daylight and rarely are they seen. Daytime meteors are so uncommon that many people erroneously think that they can only be seen at night. The Beta Taurids are never newsworthy.

Radiants can look very effective in their composite action photos or diagrams but they are not typically seen so picturesquely in the real night sky. The textbook analogy for the appearance of meteor streams is the view one has when standing on a bridge above a series of railway lines receding into the distance. The approaching trains are strictly parallel in paths but appear by perspective to open out toward the observer's point of view as they approach us at speed. The divergence is purely a line-of-sight effect as is the convergence into the distance. It is a puzzle to clearly identify the Tungus body with a known meteor stream by date and possibly by position of origin in the sky given that the TCB was entirely alone. Can it have been part of a recognized whole group? If so, where were all the others? There are no records of any other meteor events or sightings for that day but there is as strong assertion of such association if the TCB truly appeared shortly west of the Sun's position that morning.

Another incompatible feature with this hypothesis concerns the observed tail that was by all accounts bright and following the object. There was no dense, dark, smoky one seen preceding an assumed cometary fragment that would be pointed away from the Sun in orientation. This voids any clear explanation for the body of material that was dumped westward toward Europe as the result of a regular dust tail. Applied calculations suggest that there should be some chemical traces of such an exploded cometary core.

The Earth's atmosphere is an efficient but ultimately limited shield against anything solid hurled at us as well as harmful radiation from the Sun and other sources. The incoming tiny debris of cosmic rock burns up under ram pressure and friction as they enter the Earth's protective blanket of gases. Meteors may not actually flare into view until they are many degrees removed from the radiant point of apparent entry. A discerning observer may be able to trace their direction back to that radiant point under favorable conditions in a dark sky. Micrometeorites hitting are going on all the time as the minor damage to satellites sustained over long-term orbits clearly show. Conversely, a major asteroid would more likely fragment under the stresses of atmospheric entry depending on that wide range of parameters. For this reason alone, the TCB must have been a relatively dense and cohesive entity, compared to an average space rock of which we have cabinets full of specimens.

Even in daylight it is remarkable that no other pieces were large enough to be seen or heard to reach the ground that day. There is broad agreement that it was a comparatively small and strictly single body. If it was the nucleus or unusual coma of a comet, it was remarkably small as such things go — typically several miles in diameter. A more likely suggestion is that it was a kind of fragment rather than a discrete and complete cometary core. A monolith of cometary ices protected by a larger stony core and dark crust penetrating and exploding in one instant is a model. The complete absence of any observations of its approach whilst still thousands of miles removed remains unexplained. The standard image of the core of Halley's Comet imaged from the Giotto probe is an awesome snapshot, one of the best ever in planetary science. It is notably larger than the TCB is thought to have been (as are all known cometary nuclei). The present discussion hinges on trying to connect it to the Taurid Complex of meteoroids. There is nothing to suggest that Beta Taurids are significantly different to other meteor streams or that they harbor any anomalies. They are, after all, a minor group.

At the time of writing the unusual comet P/2010 A2 has appeared in the asteroid belt. It can neither be termed a comet or asteroid and whilst the convenient term "centaur" presents itself, this new body bears a peculiar X-shaped filamentary nucleus. That it is an asteroid whose rotation has been spun up to the point of shedding mass appears as a possible explanation. A collision between two such bodies in another. Returning to known comets, the nucleus of Comet Borrelly as revealed in a flyby by Deep Space 1 in 2001, is a typical five miles in size and a very dark body. Gazing at the shots taken from 1,500 miles away, a member of the imaging team exclaimed that "these pictures have told us that comet nuclei are far more complex than we ever imagined. They have rugged terrain, smooth rolling plains, deep fractures and very, very dark material."

Contrast that to the Saturnian moon Enceladus, whose face is permanently brighter than newly laid snow, with a 90% reflectivity of light. These hands-on and revealing moments promote space research as high adventure.

Scientifically, we must guard against over scrutinizing what little data there is on the appearance and path of the Tunguska body. From the timings and their locations we could sketch out a brilliant roundish body first seen at a height of 50 miles travelling NW and arriving at a 30º angle to the ground at 90,000 miles an hour or 25 miles a second. These are, of course, tentative values, for the sake of discussion. A projectile covering 400 miles in 10 minutes is moving at a much lower rate of 0.7 miles a second. That would assume that it first became visible 400 ground miles SE of the fall zone. These are all major assumptions. At what height and over which latitude and longitude did the TCB first become visible to the naked eye is a question that goes unanswered.

Anyone can play around with some basic trigonometry involving height, distance covered, angle of approach and speeds resulting from the gross assumption of time elapsed and a point of entry. If one constructs a right angled triangle with the TCB placed an arbitrary height above the ground and an assumed distance from that ground point to the impact site, the hypotenuse of the triangle represents the flight path of the object. If a low angle of approach is deduced from the interpreted observations and an approximate five minute period of observed flight assumed, then we can calculate a rough estimate of the speed of approach. This method is actually too flexible to provide a useful measure, given the lack of reliable values, but it is an exercise in how such things might be calculated.

$s = d/t$

Where s is speed, d is the distance travelled and t is the time taken.

If you imagine an object with a 10, 20 or 30 degree angle of approach over 600 ground miles and allow 5 minutes to cover the downward slope from above, it is moving at an approximate and averaged 2 miles a second or somewhere over 7,000 miles an hour on its path. For the purposes of discussion, this seems like an acceptable picture.

The Taurid Complex is possibly the largest patch of material known in the inner solar system. Comet Encke and the Taurids have deteriorated from probably a much larger comet in the past 20,000–30,000 years. There is evidence from the lunar surface studies of micro craters that the incidence of interplanetary dust in the early part of that very recent era is greater than the contemporary level by a factor of ten. This is corroborated by Arctic ice core samples for the dust settling on the Earth during the same period. If we stretch the implications, we might propose that tiny pieces of Comet Encke have indirectly hit the Earth in the past.

Observed Beta Taurids on other annual encounters have been tracked through the atmosphere at about 17 miles a second. The cores of the Taurid streams are

likely to be composed of weightier material than the tiny rock pieces that generally constitute meteors. IRAS, the Infrared Astronomical Satellite, had observed
this during a mission of the 1980s. There is no suggestion of any anomalous presence or special feature deep within them. We pass through a periphery of the
Taurid Complex in their current position and the precession of their orbit is estimated to take approximately 7,000 years. The Earth is not therefore encountering
the core of a great stream in this epoch. For a fuller discussion of the effects of the
past, look up "Coherent Catastrophism".

The material from a fall in Farmington, Kansas on 25 June 1890 proved to
be among the youngest of known meteorites. The techniques of study have improved enormously since then. By its space exposure age, it has been deuced that
it had a previous free orbit lasting a mere 25,000 years, less than 10% the age of
the next youngest meteorite specimen. Both the date of the fall and the observed
radiant position indicate that it was a Beta Taurid.

It is thought that the greater Taurid's cycle of activity peaks approximately
every 2500–3000 years when that core passes closest to Earth, potentially resulting in far more populous showers. Earth may have passed through the tail of
Comet Encke around the year 1 AD and the next peak may be due in about year
3000. There is no powerful pointer to 1908 or the early 20th century as some cli-
mactic point for the Taurid Complex interacting with the Earth's path, and we
have little basis for advocating that the TCB was a peculiar Beta Taurid meteor.

Obviously, there is an orbital period for the Taurid Complex cores. If its orbital node evolves to the position of 1 AU, a sustained bombardment could result.
The path of the Earth and that of the Taurids would intercept for a few days or
weeks. Two other sets of four lesser showers are related to the Taurids known as
the Chi Orionid and Piscid streams.

Another paradox in this version is that a proposed piece of cometary material
was somehow robust or protected enough to survive downward within 5 miles
of the Earth's surface, then abruptly incinerated itself entirely in a hot explosion
decisively not caused by impacting the ground. This dénouement still applies to
a high-speed cometary TCB entering the atmosphere at a hypothetically steep
angle to the horizon in daylight, a description less favored overall.

In the absence of reliable data on the body's size, speed and angle of approach,
we are left to envision, for instance, an outer crust and icy layering initially pro-
tecting but then ablating. On entering and descending through the Earth's atmosphere this natural shield was steadily stripped away. This clearly could give rise

to the glowing head and fiery trails generally described. No smoky tail was ever reported, but rather a bright and iridescent one by all accounts. Ablation is the loss of surface material from a meteor or spacecraft through evaporation or melting due to friction with the atmosphere. Scientists learned a lot on the subject whilst designing heat shields for spacecraft.

A St Petersburg group is recently quoted as suggesting a comet entering Earth orbit and slowly spiraling down. With a perigee over Antartica and a trajectory changing to settle moving northward along longitude 101º, it gradually lost altitude and speed over a four-day period. It finally decelerated and caved to gravitation over central Siberia. The erudite names of Nilkolsky, Schultz, Schnitke, Tsynbal and Medvedev attach to this viable proposal and due credit must be given. Unfortunately, there is no observational evidence to underpin the concept. It could be associated with the variations in the geomagnetic field recorded three times prior to the fall and all in the same locale westward in Europe. The "Weber effect" and the inexplicable magnetic precursor to the event could suggest an elliptical orbit with a perigee over Kiel rather than an object zeroing in on Germany or that region of Europe in approach.

Current ideas about comets are largely based on the American Fred Whipple's "dirty snowball" hypothesis from the early 1950s. (Asteroid 1940 Whipple is named in commemoration of the first astronomer to successfully describe them.) A concentration of frozen hydrocarbons and loosely bound rock in an icy conglomerate form their basic structure. Generally consisting of a nucleus and surrounding coma, they temporarily develop ion and/or dust tails when heated and excited by proximity to the Sun and the strengthening solar wind now closer to it. Comet tails are pointed away from the Sun by radiation pressure, and are not directly dragged behind the head. Thus such tails have been termed interplanetary windsocks on a grand scale as the comets activate and shed material in response to heating by the Sun, the solar wind and radiation pressure. Tails have been observed actually preceding comets in retreat from the Sun. The reports concerning the appearance of the body we are studying do not fully fit that of a comet or a piece thereof.

There is a saying that a comet is the closest thing to nothing that can still be something. Much of their further tails are illuminated vacuum. They may ultimately be the very commonest objects of all in the solar system.

Compositionally, a range of organic compounds including formaldehyde, hydrogen cyanide and methyl cyanide accompanies the water ice, frozen methane, ammonia and carbon monoxide components. The former possibly formed in interstellar space and were incorporated into comets in the early development of the solar nebula. They remain lodged in the Oort Cloud in great profusion of potential comets waiting to be lured sunward from distances like a major fraction of a light year. The sheer quantities of potential comets waiting there are assessed in trillions. The Kuiper Belt could be another source for them.

About thirty a year are discovered or recovered. When a prospectively new one is found, the International Astronomical Union assigns an interim designation of the discovery year followed by a lowercase letter in the order of discovery for the year. The name of the discoverer(s) precedes the designation. When a reliable orbit is established — which may take some intense work in celestial mechanics — the comet is given a permanent designation followed by the roman numeral that is the order of perihelion passage of observed comets that year. If the comet is found to be periodic, the letter P followed by the discoverer's name is used. Comet Bennett 1970 II was just passing through. Comet 1910 P/Halley is an old returning friend expected again in the year 2062. Its last visits were in 1986 and previously in 1910. Edmond Halley deduced that sightings similarly described in 1456, 1531, 1607 and 1682 were one and the same.

Comet Hale-Bopp C/1995 01 was probably the best seen of the last century. It actually passed closest to Earth in March 1997 at a distance of 1.315 AU.

The ancient Chinese term "broom" or "guest" star is certainly evocative in appearance but the general sense associated with the appearance of a comet traditionally was that of gloom and doom. Genghis Khan apparently drew inspiration to conquest in 1222 from the sight of one and the late Julius Caesar's spirit ascending to heaven in characteristic deportment ceremonially took place in 44 BCE. Halley's Comet formed a momentously bad omen in 1066 for the Anglo Saxon King Harold of England who was shortly swept from power and his kingdom usurped by William the Conqueror. It is directly depicted on the Bayeux tapestry.

Equipped, then, with a grasp of comets and atmospheric physics, we see that the notion of an inner core or nucleus component reaching as close as a few miles of the surface and then detonating so completely does not fully add up. This lowers any confidence in a cometary model. There were no obvious pieces of the TCB left, nor any cratering formed by the event or along the path of the projectile. It is often referred to as an impact but there was no collision with the ground of any material object. Kulik and company found this profoundly mysterious. In many ways we still do. Those four smaller radial patterns in the devastated trees must indicate four actual explosions felling them. The larger pattern continuing outwards suggests a final grand bang. This supposes that the body was breaking up at the very end of its career. Was that fragmentation before the explosions or the result of the blasts? A comet of low density that decelerated in the atmosphere then exploded does not fit the dynamics model.

The testimony of Semenov at Vanavara does provide one small objective clue. He stated that "suddenly and directly to the north, over Onkaul's Tunguska road, the sky split in two and fire appeared high and wide over the forest." The witness can be placed on the exact spot in Vanavara with the object at 50 degrees elevation from his reference points that morning. He went on to say that "when the sky opened up, the hot wind raced between the houses like from cannons which left traces in the ground like pathways and it damaged some crops." A full analysis of this witness and his family's testimonies indicate a column of fire 12 miles high and 1 mile wide. In the murky world of statistics for the path or any structured measurements of the TCB, this is very encouraging. It is the best description that can be built from reliable observation. A neighbor is on record as stating that he felt as though he had been seized by the heat and that there was a great clap of thunder and damage to his house. There at Vanavara the heat of passage was clearly felt with the major shock waves knocking Semenov from his porch. His daughter gave a similar testimony of being scared of the great sounds as fire brighter than the Sun poured from the sky to the north.

The nature of the TCB is still unproven and we will briefly explore some possibilities. No one has been able to construct a fully viable and accurate model despite some enterprising attempts. To this day there are too many unknown variables, insufficient facts and unexplained features. For example, no reports refer to the object splitting directly or fragmenting in flight or material dropping from the sky at any point. Neither is it fully or reliably possible to reconstruct the visual and audible accounts into a coherent whole scenario. The Hungarian Academy of Sciences has an intriguing simulation of the sounds of the Tunguska event on their website.

The Academician Fesenkov hypothesized a case requiring a cometary body five times the density of water and 240 feet in diameter coming through at 37 miles a second. With those dimensions it was notably small and more likely a fragment of a greater body continuing to orbit the Sun. His results received some coverage in the *New York Times*, including the estimated 1,000,000-ton mass for

the body. Other cases and figures are equally possible to construct. Many propose a lower speed whilst generally in favor of a small body. We can speculate
on that incandescence appearing at about 50–80 miles altitude according to normal behavior for familiar meteors on their downward paths. They generally flare
briefly into view where the density of the atmosphere begins to provide sufficient
friction. We may, with extreme caution, interpret the witness accounts that it
was first seen at a height equal to 50 miles. Fesenkov's case assumed a high comet
density; they are averaged as much closer to water. Hughes and Brown, two British astronomers concurred with a cometary model during the early 1960s.

V. Bronshten of the Russian Academy of Sciences undertook some serious
work on the sky glows following the explosion. He suggested that the outer part
of a comet had its dust conveyed westward by the Earth's gravitational field. This
would explain, by the scattering of sunlight, the white nights experienced over
western Russia, Europe and the British Isles. The winds of those dates were also
conducive.

The thermal flash descending on the forest was several thousand degrees in
temperature and the detonation took place between 3 and 8 miles up. These figures for the altitude of the explosion are, at least, corroborated and reliable facts.
That the heat flash was momentary and immediately extinguished by the ballistic
wave is a reasonable but endlessly debatable conclusion. Exactly how long that
prodigious pulse of heat and light endured remains unknown. The light flash
probably lasted for several seconds before being extinguished but here we run up
against other crucial unknown quantities like the height and the temperatures
associated.

A huge cometary dust tail dissipating in the atmosphere over a large area of
sky would go a long way to explain the many white nights over Russia and Europe to the west that immediately followed. Some induced aurorae may possibly
have contributed a busy ionosphere above the accompanying high, bright clouds.
A final trajectory for the TCB coming from the SE and moving NW is agreed.
That a massive quantity of some substance was thrown out westward and in the
general direction of Europe is a simple fact of record. To the other points of the
compass there were very few observers and no known reports. Perhaps it could
have been heard in the Arctic Circle, if anyone had been there. There is one small
adjunct that in 1941, a nomad named Chardu was recorded as stating that from
the trading station at Essey, 530 miles north of Vanavara, he had heard distant

rumbles and noticed strong winds springing up over the tundra. Regrettably, there is an intellectual vacuum along the trail of Tunguska.

No descriptions of the TCB suggest that there were any significant dust tails seen on the way in. Conversely, it was noted to have a blazing bright tail, possibly quite colorful. There is no indication of a dusty or smoky tail following and certainly not preceding the body. That would have better described a true comet but even allowing that it came out of the Sun in direction, no cometary tail of ion or dusty nature fits the reports.

Those naked eye accounts are all we have to go on. They are not the objective results of trained scientific observers by any means. The TCB was outside the sphere of experience of the average Tunguski nomad or the mainstream of human experience.

Total ablation of the TCB took place at 15,000 degrees C according to the work of V. Svetson in 1996. The explosion broke the body into a host of small fragments mere inches in size with a temperature building up high enough to entirely melt them. The microscopic debris was then condensed in the cooling atmosphere and scattered very widely. The theory favors a small stony asteroid as the source body. A carbonaceous one would have broken up higher and a metallic meteor would have partly survived to hit the ground in hot large pieces. A high density iron projectile would more likely have reached to the ground as a single unit. The figure for the temperature of the actual explosion is generally lower in estimate, closer to 3,500º C. In the theory of dynamics applied to a body entering the atmosphere, the size range of 30–330 feet is required for a stony meteorite to disruptively explode several miles above ground. Such a meteor, either smaller or monstrously larger, could hypothetically survive to the surface in part. The aerodynamic forces would probably fragment the larger meteor by stress whilst in flight. The scenarios are different for the very small and very large objects in transit. It seems unlikely that any asteroid affected by gravity and atmospheric friction alone could have delivered the punch that remains to be accounted for. The discussion remains open and some interior source of energy is increasingly evident in the real scenario.

There are plenty of tree debris and soil samples to sift through, accumulated over the decades. With improved techniques of geochemistry, one specific aim would be to one day be able to state confidently something like "the forest underwent a thermal pulse of 4000 degrees C for a 5.5 second period before a cold ballistic wave extinguished it over 10 seconds" as a conclusion. But the trail is growing cold after more than a century and no such statement can yet be made.

Recently, E. Drobyshenski of the Russian Academy of Sciences (Universe Today website 27 March 2009) has proposed a new adjunct to the cometary theory holding that a piece departed from the main body of a comet busily grazing the Earth's atmosphere and began a comparatively slow descent. At this very high and orbital altitude an explosion dispatched the main body back into space. The high hydrogen peroxide content of the descending fragment heated up and finally chemically exploded under the temperature and pressure conditions encountered at an altitude of 5 miles. Oxygen and water were formed in an explosive disassociation. This also places the energy of the chemical eruption much lower than the kinetic energy of the TCB. It was a huge act of electrolysis. Had it impacted and exploded, the energy released would have been far greater.

This further refined cometary theory of an exploding volatile-rich tiny piece embedded in a mainly dust body has its scientific good points. The powerful chemical explosion, huge dust residue and zero meteoric fragments as an after effect are rationally expounded. Oxygen strongly supports combustion but does not burn. The complete absence of observation of any parent body coming or going remains a problem. That a comet designated 2005NB 56 passed within a grossly estimated 6.2 million miles of Earth on 27 June 1908 is not certain, and neither is a predicted return in 2045. A direction of approach entirely out of the Sun and in permanent daylight is a viable supposition, but withdrawing into space without being seen in any way defies simple explanation. There are no records or an observer's lucky glimpse of a high altitude flash or explosive anomaly on the fringes of the atmosphere, either. That the TCB was passing through is not a viable proposition.

The researchers Petrov and Stulov had proposed a cometary model with an ultra low density, referred to as the cosmic snowflake model, which proved completely untenable. The body we are dealing with was obviously dense and coherent enough to mostly survive the stress of entering the atmosphere. The further suggestion that it was an archaic space body, a truly primeval piece of material left over from the proto planetary formation of the solar system, is not consistent with knowledge of regular asteroids although some minor planets are primitive. Declaring it a member of the Apollo group is the best-fit theory if one insists on an asteroid family identity. Nothing resembling a TCB lying in wait is apparent in asteroid or comet studies.

Another recent suggestion, by V. Romeiko, still favoring the Comet Encke fragment idea, holds that the predominantly negative ions in the cometary piece were suddenly exposed to the positive ions in the lower air, igniting the explosion. That the particles were very highly charged is essential to the argument. This represents the very latest Tunguska theory as mentioned in the University of Bologna press release from July 2009.

The proposition that the event was responsible for a few degrees lowering in temperature in the northern hemisphere for a few subsequent years is possible but not conclusive. Clearly, large amounts of particulates were dumped and suspended in the stratosphere by the violently explosive effects. With this material spread over a vast area of the middle atmosphere, sunlight could have been scattered and the general temperature dropped for many months. This cooling may, alternatively, have been due to other effects like volcanism in the same period. The Ksudach caldera complex had been active on Russia's Kamchatka Peninsula in March the previous year with several major eruptions affecting the atmosphere. This area lies in the far SE of Russia and was probably even more thinly populated than the Tunguska region. Siberia is so vast that Kamchatka is noted to be closer to Denver than to Moscow; it is one of the most remote areas of planet Earth.

With the Tungus event, how the explosions propagated the huge quantities of dust mainly upward and presumably westward is another part of the mystery. The initial wave impacting the Earth, partly bouncing back up to meet the main explosions and following the extinguishment of the thermal wave carrying the mass of material upward once more seems a rational description. There was a thermal and must have been a ballistic wave, but which arrived in what order to inflict what type of damage to the forest is unresolved. If we return to simplicity here, we can theorize that from the known height, the thermal wave of the fireball struck the trees initially, shortly followed by a greater ballistic wave that extinguished the blaze and perpetrated far wider damage downward.

The altitude of the explosion can be roughly corroborated by several lines of reasoning but not the strength, nature and order of the impacting waves. No thick blanket remained of meteoric debris nor was any scattered field of material left over. The ground heaved and gushed whilst the forest wilted beneath the heat and forces unleashed but there were no showers of tiny meteorites or cometary dust.

If a massive quantity of microscopic material was simply caught in suspension, the spin of the Earth itself would have conveyed it westward, even without a wind factor. This would describe the purely westward shift of the post-explosive material. The great white nights were not seen in the southern hemisphere at all. The debris was smaller than the average dust particle and the pressure waves from the initial explosions(s) went around the world twice. That they were four in number is deduced from the barographs in far-off England. It is reasonable to suppose that the detonation acted to abruptly place huge quantities of water in the upper atmosphere. That would be consistent with cometary ices abruptly liberated from their nucleus. On cooling, the water droplets froze into ice crystals swiftly dispersed by global circulation in the form of noctilucent clouds. This would be a partial explanation of the dazzling sunsets and diminished clarity of the air. The great dust residue from the explosion cannot have been drawn out of the air by the sheer localized intensity of heat. The bright nights seen to the west for weeks were widely seen but never explained. Stokes' Law describes the settling of tiny particles in a fluid medium.

The Comet Moorhouse was observed in September–October that year and in 1910 the Earth probably passed through the long tail of Halley's Comet with no ill effects or even public awareness. Its tail's length was estimated at 100,000,000 miles. The tail of the Great Comet of 1843 was probably three times longer. As for a closest approach to Earth for any recorded cometary body, Lexell's Comet of July 1770 subtended almost 2½ degrees in the sky, over four times the apparent size of the Moon. The distance was 0.0146 AU or 1,360,000 miles; five times that of the Moon. That comet, formally known as D/1770 L1, has not been seen since. The Comet Hale-Bopp 1997 was the best show of our era so far.

Hot on the trail of comets, observational kudos go to Robert H. Mcnaught of the Australian National University as the reigning champion of cometary discoveries. His record includes 25 long- and 17 short-period first time finds.

There are 3,648 known comets of which 1.500 are Kreutz Sungrazers and 200 short period objects as of early 2010. Naturally, the statistics continue to expand and surprise appearances still occur regularly. There are many Near Earth Objects. The Great Comet of 1843 was the brightest accurately and scientifically recorded of them and the Daylight Comet of 1910 the most prominent of the 20th century. It was occasionally confused with Halley's in human memory. Kulik applied himself to other meteor studies but did consider that an iron meteor companion of Comet 7/P Pons-Winnecke comprised the TCB. He wrote a brief report to the Academy in 1926 on the speculation. Pons-Winnecke made an impressive appearance in 1927, second in proximity only to Lexell's comet for a close approach to Earth.

The small asteroid MN4 is expected to pass 18,600 miles from Earth on 13 April 2029. This is closer than some geosynchronous satellites in their orbits and it will be affected by the Earth's gravitational pull. It should be about 3rd magnitude in brightness and therefore within naked eye visibility with an appreciable angular shift in the sky over a few hours. It should be a medium-bright point source in the night sky and will actually traverse between the Earth and the Moon. Enumerated and named as 99942 Apophis, it has been termed the world's most threatening asteroid since its discovery in June 2004. The record for close approach stands at 4,000 miles for the 20 feet meteor 2004 FH detected hours before passing 1/60 of the Moon's distance.

The traditional record of closest approach by an asteroid of significant size is held by the S type Apollo asteroid 69230 Hermes, passing approximately 450,000 miles from Earth in October 1937. It was moving at 5 degrees an hour but at 8th magnitude was not visible to the naked eye. Hermes is about 300 yards in size. Subsequent radar work in 2003 revealed not only that is a binary body of those estimated equal sizes but that it had passed even closer in 1942, entirely unobserved at the time.

Ever-smaller asteroids have certainly passed nearer since and continue to do so. The asteroid 4179 Toutatis was contacted by radar in 1992 at the close point of its chaotic 4-year orbit, passing about 2,000,000 miles and less than one million in 2004. The plane of its orbit is closer to the plane of Earth's orbit than any known asteroid of significant size. The radar studies gave a reliable shape of 1.70 by 2.03 by 4.26 +/- 0.08 kms in dimensions and an impressive image of its rotating shape. This one looked more like a dumbbell than a potato.

A glance at the NEO Program website lists at least ten close approaches a month — depending on the definition of "close." The term "miss distance" is given in both Astronomical Units and Lunar Distances. For 30 June 2009 two very minor bodies were estimated to miss us by 49.7 and 9.6 LDs.

Finally, the division between a rocky asteroid and gaseous, icy comet is not entirely strict. The asteroid 2060 Chiron discovered in 1977 is a truly far-flung object. It displayed a coma and tail on its perihelion in the 80s and is one of several asteroids following cometary-like orbits. Conversely, 5335 Damocles is likely an extinct and inert comet rather than as asteroid in origin. Here in the early 21st century it is considered that all regular NEOs greater than an estimated 9 miles in size have been discovered by now. But the skypub archive from November 1996 contains a revealing paper entitled, "Are They Comets or Asteroids," by Stuart Goldman. Following up on a possible new comet using the I meter Schmidt telescope in Chile of a tail-sporting, 18th magnitude object:

> Subsequent observations suggest that a narrow, straight tail of Comet Elst-Pizarro (P/1996 N2) may be recent—perhaps the result of dust emission between late May and early July. However, not everyone thinks the object is a comet. David Balam (University of Victoria) notes that it has no coma, so we might be seeing the effects of debris scattered by an asteroid collision.

It was been suggested that 5–10% of NEOs are really inactive comets inaccurately assumed to be asteroids. Their nuclei are used up and quiescent now, wandering planetary space. They cannot be put forward as a TCB candidate. Statistically, an asteroid is a numerically greater chance.

A useful unit in planetary astronomy is the Astronomical Unit. It is equal to the average distance between the Earth and the Sun and approximates to 93 million miles or 150 million kilometers. Taking the mean distance of Pluto, varying between 39.48 and 49.31 AU from the Sun, the solar system is therefore just over 6 light hours in radius. It is 4.3 light years to the closest stellar neighbor; the Alpha Centauri system is farther by a factor of over 6,000.

Cometary hypotheses for Tunguska have some strengths and an enormous weakness in failing to account for how could it have descended without already breaking up and entirely vaporizing. Let us review a plausible rendition from Norton's "Rocks From Space" (p. 97), then consider a particular comet candidate.

> Perhaps on that June morning in 1908, Earth had encountered a comet: a body composed of a mixture of ice and metallic and silicate chunks. The comet entered the atmosphere at the low angle of only 17 degrees. It travelled for 400 miles in a south-southeast to north-northwest direction, beginning above the north shore of Lake Baikal. The initial mass of the body was estimated to be about 1 million tons. It was travelling between 17 and 24 miles a second when it hit the upper atmosphere and it retained most of its cosmic velocity throughout its trajectory. It lost nearly 95% of its mass in transit so it was probably between 20,000 and 70,000 tons when it exploded about 5 miles above Tunguska.

It paints a reasonably convincing picture. We applaud a theoretician who can come up with justified specific values and a plausible description. It is consistent and coherent enough to put forward. The majority of the energy was transferred into a shock wave that struck the forest beneath. The trees immediately below presented a smaller cross section to the shock wave and were able to remain standing although strongly singed by heat and mostly killed in the process. This formed the 5-mile telegraph pole epicenter. There is the resemblance to Hiroshima ground zero with no suggestion of an atomic explosion although major hot shock wave(s) impacted the ground.

Other scenarios exclude the ballistic wave felling even a single tree and attribute that all to the blast wave with the ballistic one quelling the intense fire in very short order. The fire may have raged for mere seconds; many trees were burned on one side only and very briefly exposed to the heat. The TCB retarded its speed appreciably, a factor affecting the downward path of normal meteors. Some are so slowed that hot little rocks practically free fall out of the sky at the finish of their trajectories. In conclusion, the ice vaporized, the heat strongly scorched the trees and a large quantity of sub micro meteoric particles was blasted upward and somehow westerly into the upper atmosphere. A reverberating shock wave could suffice to have propelled them upward.

One can see why the cometary hypothesis is most favored. Let us now consider a specific candidate that accords with the Taurid Complex and a specific range of dates for encountering the Earth in space.

CHAPTER 5. COMET ENCKE

Pierre Mechain (1744–1804) originally discovered Comet Encke in 1786. Following some monumental mathematical work by Johann Encke (1791–1865), it was shown in 1818 to be periodic and became the second comet proven to reappear time after time. This explains the formal designation Comet 2/P Encke.

Encke the astronomer successfully predicted a return in 1822. Three years earlier he had published his work in the journal *Correspondence Astronomique*. In more recent research, its nucleus is estimated to cover nearly 3 miles and the contemporary date of perihelion is a very reliable 6 August 2010.

Comet Encke has an orbital period of a short 3.3 years and was not particularly close at the time of the Tungus Event. This is an important starting point for this precise comet or comet fragment theory to gather any credible momentum and make its explanatory impact.

Comet Encke was receding from the Earth's position around the Sun in June 1908. It follows a normal prograde direction and had been substantially closer in 1904. There are no sightings or hints of any approaching body until the TCB suddenly arrived in all its drama, and this member of the solar system could be immediately excluded on those grounds. It is still there in a swift solar orbit. It was never observed losing a piece of itself. Comet Encke is, however, one definitive parent body of the Taurid Complex whilst other objects may also have contributed to its reservoir of material revolving around the Sun over time. No known body is objectively viewed as a candidate for the TCB.

We recently considered the following questions: Was the TCB's direction of origin consistent with the Beta Taurid radiant or the Sun itself? Had it swerved or gently altered course to come from the SW or SE, heading NW during its final approach? Let us examine its last moments with the best information available, then consider the wider movements of Comet Encke.

The major line of forest devastation lies along a NW to SE axis, most probably revealing the terminal trajectory. Voznesensky early derived a trajectory from SSW from the immediate verbal accounts and seismic records. The general agreement today is a final path from SSE to NNW. It may have first become visible shortly NW of Lake Baikal and approximately 400 miles from the fall site but this is by no means certain. It may have changed course over Kezhma on the Angara River, 120 miles from the zone, then passed NE and close to Vanavara. Interested parties are not fully agreed whether it passed E or W of that nearest town along its final trajectory, 37 ground miles from the explosion. It certainly was close enough for the waves of heat to be felt and disturbingly great noise to be heard there. Kulik's work placed the trajectory to include a passage 21 miles east of the settlement. It is coincidental that the trajectory was approximately parallel to the direction of the low ground wind on 30 June. The TCB was hardly gliding downward or seen to be wind affected in its course.

A watcher at Kamenka, 280 miles SW and one of the further locales of visual testimony, reported that it "broke away from the Sun." At 7:17 am local time the Sun was about $23°$ and 4 hours above the horizon and almost due East. The morning was warm, dry and clear skied.

In detail, Comet Encke was four months past its regular perihelion at that time. That brief speedy passage closest to the Sun occurs once every 3.3 years before it heads out again into the realm of the planets. During late June 1908 it was at a distance of 1.61 AU from the Sun and 0.741 AU from Earth and receding from both. It was therefore past its most active and excited stage by summer 1908.

Nikolsky, Schultz and Medvedev, all noted members of the Academy of Sciences, have recently constructed a working model of a piece reaching here and going over four orbits of Earth, including disturbing the geomagnetic field over Kiel. It finally fragmented and after its blast wave hit a Siberian forest, burning gases ensued. This is a creative attempt to account for as many effects as possible. It places an orbital perigee near Antarctica's Mt. Erubus, the southernmost active volcano in the world, where a major aurora was seen on the day. Unfortunately, the narrative stalls immediately on the critical point that there is no known split of material from the main body of Comet Encke.

The separate orbits of the Earth and Comet Encke brought the two bodies closest to each other two months earlier on 20 April 1908 at 0.59 AU distance. That is nearly 60% of Earth's distance to the Sun and is not a close encounter in terms of the two bodies' predictable ranges from each other. The comet had passed much nearer in early November 1904. This was during its previous passage through the inner solar system at a minimum distance of 0.12 AU from Earth, passing on to perihelion in mid December 1904.

Comet Encke is relatively compact and inactive. It has the smallest aphelion, the far point of its orbit, of any known comet, a modest distance of 4.1 AU. It does not reach as far as the orbit of Jupiter before turning sunward once more. Comet Encke has presumably been tamed in movement by the gravity of the Sun and Jupiter in the long term. It is more likely to be an aged entity rather than one undergoing dramatic perturbations in its path over the last few millennia. It arrived from the Oort Cloud long ago on its first go round and was captured by the dominating gravitational field of the Sun. Its orbit is now very small and no longer highly elongated in shape (by cometary standards) — a value of 0.8471 in eccentricity. It was presumably far more massive with a wider and more eccentric orbit in the past. This is entirely consistent with cometary observation and theory. Comet Encke is therefore an old, low mass, regularized object in a fixed, brief and established journey around the Sun.

Its path is inclined to the ecliptic by 11.76 degrees and the motion is prograde about the Sun like most orbiting bodies. Long-term comets are often retrograde in direction and 10% of those are gravitationally unbound and very much just passing through. They have the speeds to avoid permanent capture and, following a path somewhat affected by the Sun's gravity and heat, delve back into interstellar space. The delineation for short or long term is that arbitrary +/- 200-year period of an orbit.

The independent small asteroid NEO 2004 TG10 may be a fragment of the Encke parent body as suggested by its orbital motion. Detached pieces do not necessarily remain close to the parent body itself over time. Some proportions, probably very low, of asteroids are expended comets as such. When close to the Sun a very few are seen to flare up like weak comets. They probably still possess a thin veneer of volatiles that can go into low activity when heated. They were previously thought to be asteroids now spending the majority of their orbits at much lower temperatures, then briefly brightening and radiating as they pass through perihelion.

There was a fortuitous chapter in the discoveries of space astronomy from the ongoing STEREO Mission (Solar Terrestrial Relations Observatory). The helioscopic telescope aboard the STEREO/A was observing Comet Encke during its perihelion in April 2007. At that juncture, a very major outburst from the solar corona was propagating a mass of electrified solar material, some of which intercepted the comet. The magnetic fields associated were probably responsible for the shearing off of its tail rather than the low impact of material from the Sun. The high speed and temperatures exerted by the gas covers such a large volume of space that it actually renders an overall pressure little more than a slight breeze. The comet probably underwent a large-scale magnetic storm. The magnetic field in the solar ejection was presumably oriented opposite to the cometary magnetic field. The tail subsequently was replaced by the ongoing shedding of gas and dust as Encke continued on its path following interference from the coronal mass ejection.

What can we make of this? A disruption of Comet Encke took place before our very space-faring electronic eyes a few years ago. However, there is no evidence based on orbital paths to suggest that a piece was zeroing in on Earth in 1908. A *Starry Mirror* press release gave the imposing headline "Solar Storm Tears Tail From Comet Encke."

The stated mission concept of dual satellites, one ahead and another behind the Earth will, an estimated four years into the mission, position the two probes almost diametrically opposite in orbit. STEREO provides unique and revolutionary views of the Sun–Earth system. The satellites trace the flow of energy from the Sun to the Earth as well as reveal the 3D structure of coronal mass ejections and helps us understand why they happen. STEREO also provides alerts for Earth directed solar ejections from its unique side viewing perspective adding to the fleet of space weather detection satellites.

Note that this upheaval in the tail of Comet Encke occurred a few years ago and, of course, nearly a century after the Tungus Event. As technology advances, we can be more certain to distinguish between the genuinely rare as opposed to a previously unobserved phenomenon. That tail dislocation event was probably genuinely uncommon. It is a rarity for a comet to be going through perihelion just as the Sun releases a CME that then collides with it. Before space-probe-borne technology, any such thing was an unseen occurrence. Only lately have we designed the means to detect something that does not happen very often.

The first recorded CME was in late 1971 but an earlier one probably coincided with the first observed solar flare in 1859. They typically reach Earth in a five-day period following their eruption from the Sun, the faster ones attaining some 300 miles a second and decelerating toward the speed of the solar wind which averages 250 miles a second. That speed is extremely variable. Slower ones are accelerated up to the velocity of the wind from the Sun.

NEOs are another case in point. They exist in profusion but further funding and research will be needed if scientists are to locate them to the higher standards and greater quantities required by Congress. There was a short lived hypothesis a few years ago proposing that an unending stream of extremely mini comets were regularly entering the Earth's atmosphere from space. The idea gathered very little support. That Tunguska involved a cometary tail that arrived first above Siberia, shortly followed by an explosive cometary nucleus that had been protected in flight, has been proposed. Unfortunately, no such appearance fits the observational statements or the dynamics test for a modeled atmospheric entry.

If Comet Encke's perihelion and most active phase had occurred shortly before a particularly close passage to Earth around 1908, dropping off a sizeable chunk in our direction, then we could be onto something. However, this is decisively not the case. Its positions during the early 20th century are known and plotted. During the perihelion two orbits previously in late August 1901 the Earth lay on the opposite side of the Sun entirely to Comet Encke. One cannot posit a section of it departing and heading off to intercept the Earth by examining the orbit simulations prior to 1908. That such an event happened but took several orbits to get here is a wild stretch. There is no evidence to suggest it, even if such a view was proposed by those academics from St. Petersburg in 2009. There is simply no such case to be made from the orbit diagrams.

Observationally, comets have been seen to split up, fail to appear, or be replaced by meteor streams. They were originally produced from the prodigious spherical Oort Cloud far beyond the orbits of the Sun's family. The Oort Cloud formed or more likely endured like a huge outer periphery of primeval material beyond the solar system about 5 billion years ago. Its inner disc lies perhaps 50,000 AU from the Sun or about a quarter of the distance to Alpha Centauri. There may be a structure of a spherical outer Oort Cloud and an inner disc-like Hills cloud.

From a reservoir in the form of water, ammonia and methane ices, a proto comet may be gravitationally prompted into an elongated dive or path towards the Sun to pass through a close point before returning to the depths. It may be merely deflected further into interstellar space. A long period orbit may be established and evolve downward over time. Cometary paths therefore tend to be highly elliptical at first and often remain so over many revolutions. They may circularize and contract over time but they are relatively short lived as active bodies by the time scales of the solar system. They are comparatively low mass. Alternatively, some must be vaporized in a self-destructive dive towards the solar fires or dissipate into streams of meteoroids remaining in paths around the Sun. Many meteors in our skies are particles of the very final outcome of these processes.

So a short-period orbit may develop over a long series of orbits. As always, the gravity of the Sun and Jupiter cast the dominant influences. All comets have perihelia in the range of 1–3 AU. They simply would not survive a closer passage and live to tell the tale. Their volatile material would entirely boil off into space.

Other short-period comets may have originated less far from home within the Kuiper Belt lying an estimated 30–50 AU from the Sun. This too is an ongoing discussion in cometary science. But the Oort Cloud seems to be the primary and practically inexhaustible source considering its sheer scale. That distance of an innermost 50,000 AU from the Sun is consistent with the aphelia of new comets. This is about 1000 times larger than the orbit of Pluto and is their presumed starting points in their long hauls toward the Sun. The distributions of the perihelia directions are close enough to the solar apex to support the Oort's cometary cloud hypothesis. So is the directivity of the aphelia to the solar antapex. In short, with improved understanding of the origin of comets from Edmond Halley up, the matter appears to be resolved. The (plural) Oort Clouds are their repository and the Sun's gravity pulls them in, assisted by other sources of deflection.

The Kuiper Belt might properly be called the Edgeworth–Kuiper Belt. Gerard Kuiper suggested such a disc in 1951 and Kenneth Edgeworth had made a similar proposal eight years previously in a British Astronomical Association paper. Frederick C. Leonard had suggested a trans-Neptunian population immediately after the discovery of Pluto.

The dwarf planet Pluto has been shown to be the least distant and notably second largest of any KBO. It is a challenge to find out much more in the short term.

A single retrograde KBO is known. One may quarrel about its status as a "dwarf" planet and dethroned classical planet, but Pluto must be seen as the foremost member of a class of bodies. As credited, this had been suggested as far a back as 1930 when Pluto was first discovered. The class of bodies to which it belongs are the furthest flung planetary members of the Sun's family.

KBOs, Trans-Neptunian Objects or the less popular "plutinos" are other newly-coined terms in the nomenclature of small solar system bodies. The first KBO was discovered in 1992. At least and at last we have some idea with the origin of Pluto and need not see it in isolation. It is anything but a lone maverick, considering the greater territory. The label "minor planets" is still used for the asteroids. There are, however, no grounds to suggest that small bodies with powerful latent internal heat sources such as the TCB exist in any asteroid family.

KBOs have extremely long periods of several centuries and some bizarre orbits of high inclination and eccentricity. The largest known "Iris" is slightly bigger than Pluto with a 557-year orbital period. The KBO object Sedna requires over 12,000 years to circumnavigate the Sun. There are probably more bodies than a single one beyond Pluto that are larger than the traditional ninth planet. This was something of a breakthrough. From speculating that an object bigger than Pluto could lie beyond its orbit to glimpsing a real entity both larger and further was an exciting development. The New Horizons mission, en route to Pluto since early 2006, has a NASA webpage with an actual moving countdown to its planned closest approach in July 2015. This is the only planet yet to be visited by a probe; we do have a single close up of Uranus and Neptune by Voyager. That sort of imagery rolling in for the first time has a singular charm.

Halley's Comet has a 76-year period and has been seen regularly since at least the 3rd century BCE, again according to the Chinese annals. After some historical and dynamical studies of previous and similar appearances, it was triumphantly predicted by Edmond Halley (1656–1742) to return on schedule in 1759. It is in orbit around the Sun and returns to the inner solar system regularly. This was a bold stroke of discovery at the time, simultaneous to Newton's theory of gravitation itself. Incidentally, Halley insisted and largely paid for his friend Newton's great thesis to be published in 1687.

Halley is probably one of the largest comets and notably has a retrograde orbit like over 50% of the long period members. It is unfortunate that its appearance in 1986 was a disappointment in that its perihelion and brightest phase were on the far side of the Sun on that visit and was not directly observable from

Earth. In our lifetime it has gone unseen at its best. The Giotto probe's imagery of its nucleus was a remarkable picture, however. Halley's discovery was a strong vindication of his associate Newton's theories but his hypothesizing in a lecture to the Royal Society in 1694, suggesting a link between the biblical flood and a cometary impact, earned the disapproval of the Church. He soon "reconsidered" his publicly stated opinions.

The French astronomer Laplace later speculated that a close encounter with a comet might raise huge and devastating tidal effects on Earth. It was a rational idea at the time, but is now ruled out by our knowledge of the very low mass of comets in relation to the mass of the Earth. Steel has calculated a convincing chance as low as 1 in 300 million for a comet actually hitting Earth, much lower than that of an asteroid strike. Purely statistically, the speeds of approach increase for asteroids, short period, and long period comets, in that ascending order, for their hypothetical encroachment on the Earth's path. A comet in the inner solar system is likely to be travelling much faster than an asteroid in a comparable position. He also puts up a convincing argument that a piece of Comet Encke comprised the TCB. Consider the entry angle, speed, point of origin and above all the date for the event. The latter two are clearly related to the Beta Taurids that descended from Comet Encke and other material sources. The previous two conditions may be inferred and are supported by the weight of evidence to indicate a low entry angle and low speed (Steel, p 181). It can be rejected as unlikely to have been from any other source other than the Taurid Complex.

As noted, the statistical chance that it was a meteor/asteroid runs far higher than a cometary body. Comets, however, have a far deeper history in human records. Asteroids have only been known for the last two centuries and are still in the domain of the astronomers.

This very thing may have happened before. A piece of the same body could previously have hit the Earth. The 2-mile dry lake Umm al Binni, lying in the Al Amarah marshes of southern Iraq, shows evidence of being a Holocene impact crater dating from approximately 8000 BCE. Any regular geological formation is ruled out. It is conceivable that an impact took place in an area that was shallowly beneath the Persian Gulf at the time and was responsible for major tsunamis and extensive flooding. It may have referred to or formed the very inspiration for the Epic of Gilgamesh or Biblical Flood that became incorporated into holy writings. Edmond Halley's retracted speculation may actually have been on the right scientific path. It could explain the sediment deposit discovered in the ancient city of Ur and the Sumerian Deluge that is interpreted as a geophysical and historical fact. Certainly there were major upheavals including climatic changes around 2200 BCE in that region.

Admirers of the Comet Encke have posited that the ancient symbol of the swastika could derive from the view of a comet from head on. The "Cosmic Serpent" book's section on the Mawangdui Silk Texts written by Clube and Napier in 1982 suggests that the Chinese Han Dynasty silk work describes the breakup of Comet Encke. Fred Whipple suggested in his work that the comet's polar axis is only 5 degrees from its orbital plane. In the past, when it was more active, it could have presented a pinwheel appearance, as is suggested in Wikipedia.

Returning to the stricter disciplines of science, the JPL Solar System Dynamics orbit diagrams show the paths of the solar system's members, great and small. They give a moving picture that is worth a thousand words.

Chapter 6. The Sun, The Cosmos and Coronal Mass Ejections

The familiar Sun does not have a hard demarcation zone of a surface. What the naked eye or simple optical telescope suggests is rather a visible surface termed the photosphere, collectively about 200 miles thick. ("Photos" is Greek for light.) This is the zone where the gaseous layers change from opaqueness to transparency.

Up from the surface, the photosphere ends and the chromosphere begins where the density of negative hydrogen ions drops too low to result in appreciable opacity. Note that one must never observe the Sun directly through binoculars, telescope or any optical aid, even with a darkened lens cover, but rather project the bright image onto a piece of white card. Clean, dustless lenses are also a must here. Adjusting the orientation of the telescope, some sunspots can usually be seen and focused on crisply. Above all, safety for the observer's eyesight means never looking directly at the Sun; that is imperative.

How hot and far away is the Sun? The surface is some 6,000 degrees K but it is a strongly heterogeneous body, with deeper internal parts very different in nature and some 15 million degrees K in temperature. Its upper atmosphere above the photosphere is divided into three ascending layers, the chromosphere, the transitional zone and the corona.

Degrees Kelvin is the absolute scale of temperature used in physics referenced to absolute zero at -273.15 º Centigrade, where thermal motion ceases. It is expressed as a single "K" in usage. The distance is by definition one Astronomical

Unit equal to the mean Earth–Sun distance of 93.0 million miles or 149.6 million kilometers.

A simplified description of the structure of the Sun from the inside out would be thus: A nuclear burning core overlaid by radiative zones reaching upward toward a three-tiered convection zone in turn leading up to the visible photosphere. Above that hangs a permanent huge ceiling of a hotly rarified and busy corona extending outward into space in all directions. The release of energy is beyond human comprehension. The pressure at the core is perhaps 340 billion times that of air pressure at sea level.

The Sun is an average star of spectral type G2 V by the spectroscopic classification of stars. Breaking down the wavelengths of a star's spectrum of colors provides enormous information about its character, both physical and chemical. Occupying a central position on the stellar Main Sequence, it is in early middle age. Approximately 5 billion years have elapsed since it ignited and its nuclear fires began to glow. The Hertzprung–Russell or HR diagram plots types of stars, and the Sun duly occupies the middle section for many such properties. It is entirely average in spectral class and luminosity. (These are the two axes for plotting stars and they reveal their many types in a methodical great chart.) This HR diagram is both a snapshot of the present state of a large number of stars and the pattern for their generalized evolutionary paths. The Sun's mass, volume, density, composition, temperature, luminosity and other statistics are therefore entirely standard in the hierarchy of stars. Stellar types fit in well with the scheme of this major scatter graph that formally plots absolute magnitudes or luminosity against their spectral types. Much can be seen and inferred on types of stars and their evolutionary tracts.

Our local star, about which the entire solar system physically revolves in a wide range of periods, is moving through its lifecycle. The Sun is merely one of about 100,000 million in the local galaxy. Assessing the great quantities of brown and red dwarves as individual stars, that is a serious underestimate of the total stellar population of the Milky Way. The Sun is nearer than the closest next star system, Alpha Centauri, by a factor of 270,000. It is the difference of eight minutes at the speed of light and just over four years. Its visual magnitude is a supreme -27.

In physical scale the Sun is 865,000 miles in diameter and the equivalent size of 109 Earths stretched in line across its face. Disturbances such as sunspots and flares visible on its surface can be larger than our whole planet. It converts hydrogen into helium by nuclear fusion processes and presently consumes some 4 million tons of material a second.

The processes convert hydrogen to helium and other heavy elements with a huge release of energy. The Sun is by far the most massive body and biggest gravitational field in the solar system, containing an estimated 99.86% of the mass of the entire system. Its mass is greater than that of the Earth by a factor of 332,830. At its equator it spins over a period of 24 days 6 hours and slower across its latitudes up to its poles of spin where the rotation takes 33 days 12 hours. Its axial inclination is just over 7 degrees and its average density is about 1½ times that of water. In mass, the Sun comprises 71% hydrogen, 23% helium and the remainder trace amounts of other elements such as oxygen carbon and others. Its gravity is greater than that of the Earth's by a factor of 28 and its volume is 1,304,000 larger. There was a crisis in astrophysics up to the late 19th century and until the first hints of the structure of the atom and the existence of nuclear reactions were discovered. Until then, no one could explain what kept the Sun going.

For the inner structure of the Sun, the consensus is that above the inner and outer cores lies a radiation zone that gives way to the convection zones. Here the less dense hotter gases rise toward the surfaces where they subsequently cool and return to the depths of the convection zone again.

The distance of the Earth in its orbit varies by a small amount. The Earth's orbit is a slight ellipse; it deviates very slightly from the circular shape. The value for eccentricity e is 0.167. The semi minor axis is 99.986% of the semi major axis from the close point of perihelion occurring in early January and the far point of aphelion in early July. The values are 91 and 94 million miles, a variation of 3%. This is another natural truth unsuspected by those ancient cultures that thought about such things objectively.

It was the Copernican Revolution that relegated the Earth from the center of all things and instead recognized that the planets, including Earth, revolve around the Sun. The detail that the Sun's distance varies slightly due to the slight eccentricity of the Earth's orbit remains unfamiliar to the common man. These conditions are, however, sufficient to speed up the motion of the Moon in its orbit around Earth during late December to early January and provide a slightly briefer lunar month at that time of the year. The whole Earth–Moon system is moving a little faster in this period around their common centers of gravity. The assembly also moves a little swifter around the Sun at that perihelion point.

Of the four small rocky planets of the inner solar system (Mercury, Venus, Earth and Mars), Earth is marginally the largest over Venus but the dense Cytherean atmosphere and heat of Venus means that only in size is it comparable to Earth. All three have low magnetic fields compared to Earth.

Mercury has the most eccentric orbit of the inner planets, being 52% further from the Sun at aphelion than perihelion. It also has a day longer than its year as a result of a slow axial spin taking nearly 59 Earth days. Meanwhile, it rapidly orbits the Sun in only 88 Earth days in an eccentrically shaped path. With Mercury $e = 0.2056$ and for Mars $e = 0.0934$. The planet Venus' eccentricity is even closer to a true circle than our own orbital path's shape. For Venus, $e = 0.0068$.

For a time it was thought that the planet Mercury exhibited captured rotation similar to the Moon's revolution about Earth, keeping the same face turned to its primary. This was proved not to be the case. The axial spin relationship is 2:3 of the year, allowing a day to night or diurnal cycle equal to 176 Earth days in length. Therefore, Mercury has a longer day than its year. Sunrise to sunset is longer than the time taken to orbit the Sun. A further anomaly is that as the huge Sun slowly crosses the sky as seen from the surface of Mercury, it briefly halts and retrogresses and then recovers to continue on its way. Imagine the sight of the giant Sun when that occurs close to a horizon at sunrise or sunset. It sets, rises again in reverse and sets once more. It can also rise, fall back beneath the horizon then rise again on the eastern horizon. Seen from the parched surface of Mercury, the Sun that is so close and circumnavigated in a rapid 88 days appears to move very slowly across the sky. It regularly retrogresses, then recovers. Mercury has the most inclined orbit of any classical planet and its proximity to the Sun creates some unusual effects.

The slow and retrograde spin of Venus on its axis is another unexpounded circumstance. Gravitational braking from the Sun appears to tidally retard the low pace of spin for both Mercury and Venus. That seems relatively clear and simple. The proximity of the Sun probably prevented any moons taking up residence there in stable or permanent orbits. It is possible that Venus originally spun in the same direction as the other inner planets but was somehow massively flipped on its axis. Such a later and major adjustment is more likely than Venus receiving a retrograde spin at birth, but this phenomenon remains a puzzle.

Earth, as mentioned above, is marginally the largest of the inner planets. With Venus, the dense carbon dioxide atmosphere and the runaway greenhouse effect allows very hot surface conditions. The remark that Venus is Hell, referring to a middle cloud layer with winds of 450 miles an hour and occasional sulfuric acid precipitation, is poetically justified.

The vast bodies of free water are peculiar to planet Earth. Temperature is an important factor that allows for water. It is a compound with some unusual traits and is very rare in liquid form other than here, and the Earth's surface is exceptional in having more water than dry land. Humans ourselves are largely made up of water.

The Sun is the source of almost all energy in its empire. The trapping of primordial heat, radiogenic heat release and tidal heating are real forces but the by far the great blaze of energy broadcast across the solar system emanates from the Sun's fires.

One distant day it will cool and swell into a bloated red giant and the inner solar system, including the worlds of Mercury, Venus and Earth, will be within the body of the Sun or in such a hot zone as to be vaporized. This is the ultimate fate of in our immediate planetary neighborhood in several billion years time, failing any other catastrophe or intervention. What changes this may trigger in the great gas giant planets as a major function of temperature or what shape their orbits may adjust to can only be speculated upon. It can be suggested that the gravitational field will be more spread out over a greater volume of an enlarged and reddened Sun. The quartet of Jupiter, Saturn, Uranus and Neptune may shift even further outward and not appreciably be heated up during the coming red giant phase.

Jupiter is a laboratory offering other insights. For a time, it was commonly assumed that the Sun was the only source of heat in the planetary system. At huge distances, great cold unavoidably results, with no exceptions, it was thought. The inner zones have the monopoly on warmth, it was reasoned. Then came some glimmerings about radiogenic or gravitational heating as an energy source and the trapping of primordial heat. All are at work in the system of Jupiter and its 63 known moons. (Radiogenic heat drives the tectonic plates of the Earth that is also unique here.) Some major surprises were in store such as the busy volcanic nature of Jupiter's moon Io. For example, Loki is the most dynamic volcano known.

Meanwhile, back in ancient Greece, Aristotle's pupil Theophrastus saw spots on the Sun's surface with his naked eye and challenged his teacher's concept of a perfect and unchanging sphere of fire. That formalized view was unquestioningly handed down to European civilization. Ancient Chinese annals as early as the 8th century BCE refer to sunspots whilst Galileo and others all "discovered" them in the early 1600s, equipped with the new "optik tube" invention. Optically, it is possible for the eye to detect those spots about three times larger than Earth when cloud or haze suitably reduce the basic glare. Galileo also found the first evidence that the Sun rotated on its axis by watching the disc of the Sun for a few weeks. This is something amateur astronomers can view with a telescope over a series of sunny days.

Gamma ray bursts are another development in observational cosmology, first discovered in connection with the Vela satellites in the late 1960s. It remains an extremely remote possibility that a major GRB could erupt in this vicinity of the galaxy. In principle, this could render a strong pulse of energy for long minutes or hours sufficient to boil away much of the atmosphere and hydrosphere whilst incinerating the Earth's surface at a rapid pace. GRBs form a narrow beam of intense energy released following a supernova event as a massive star finally collapses, possibly to become a black hole.

Here is another terminal scenario for humanity that has only been known in recent times. J.S. Mill said something about greater knowledge bringing more acute mental suffering. It is a prime example of something that has been going on since the Big Bang but has only entered human comprehension lately. Our ancestors never worried about Gamma rays.

Whilst it is clear that some major asteroid impact will inevitably occur again sometime in the coming centuries, by sheer law of averages, a local GRB is improbable. A close by release of energy in a few hours at least equivalent to the Sun's 10 billion year total output would, of course, annihilate planet Earth. It is calculated that any GRB within a thousand light years would subject us to energy 500 times the power of the Sun for a while.

A new distance record was newly set with GRB 090423. It was observed in April 2009 but not analyzed and recognized for some months as the most distant object seen to date. This happens occasionally and very excitingly in astronomy. Look up the "Pillars of Creation" from the Eagle Nebula or the Hubble Deep Field imagery from the Hubble Space Telescope for the clarity and magnificence that adorns pictures from deep space.

With a red shift of 8.1, that GRB must have occurred 13.1 billion years ago, an impressive 690 million years after the Big Bang. For those of us who like the plumb the cosmological great depths this was only shortly after the "Dark Ages" and could conceivably be the explosive signature of a Population III star. These hypothetical very original stars are thought to have been both more luminous and massive than the later Population II or Population I bodies that formed later. Those early supermassive stars were uncontaminated by the heavy elements that were yet to be synthesized in an active stellar interior. There are grounds to believe that their GRBs were more intense and shorter lived in those fundamentally earlier epochs of the cosmos.

Considering the Sun, it will be about twice its current age of 5 billion years before expanding into the red giant phase if it runs true to form for stellar lifecycles. Subsequent to the red giant phase it will dramatically contract to a white dwarf in the September of its stellar years.

The Sun's chromosphere is a narrow region between the photosphere and corona, the uppermost level of the Sun's atmosphere. The temperature steadily climbs in increments and the composition and density changes greatly up to the base of the corona. This is approximately 6,000 miles above the photosphere. Here lies a major transition zone where temperatures climb hugely to about 1 million degrees K. All the visible light phenomenon of the Sun such as sunspots, spectra and Fraunhofer lines are features of the photosphere.

The huge rising convection cells of hot gas welling up from beneath the surfaces cause its granulated, bubbling texture. The lower chromosphere is fairly continuous but the upper parts are characterized by jagged spicules, short lived narrow jets of gas some 5,000 miles long clustering at the edges of super granulation cells. Spicules act as gaseous tongues and frothy jets above the bursting granules. From the upper reaches of the photosphere they fall back in minutes, rising at 10–15 miles a second up to heights of 5,000–6,000 miles. Other important features of the chromosphere include plages, prominences, filaments and the chromospheric network. Outward in the corona temperatures reach up to 4 million degrees K but the material is very thin. The density of the Sun's atmosphere drops off swiftly with ascending height. The estimate is 55 miles height bringing a decrease of 50% in density.

Coronal mass ejections are the most spectacular of blazing eruptions from the Sun. The corona is the tenuous uppermost level of the atmosphere consisting of extremely hot matter extending for millions of miles. These gigantic magnetic bubbles were formally seen for the first time in the early 1970s by spaceborne instruments. Their material is plasma, mostly constituted of helium, oxygen and other heavy elements. Plasma is the fourth state of matter beyond solid, liquid and gas. When electrons are stripped from gaseous atoms, a hot soup of ions and electrons move about in random thermal motion. The corona is far more elevated in temperature than the solar surface but is notably far more tenuous. During the eclipse of 1860, the German observer E.W.L. Tempel successfully sketched a transient bubble in the corona, a drawing that did not make any impression at the time. Just over a century later, during a manned Skylab observation, bubbles were observed at a measured average of 1 every 100 hours. In retrospect, it is seen that Tempel must be honored with the original discovery of the CME.

The corona is therefore the most energetic reach of the outermost structure of the Sun's activity. CMEs can occur anytime in the 11-year solar cycle but increase in frequency during the solar maximum. It is a fact of local cosmic nature that the Sun is a slightly variable star. This was first revealed in the study of sunspot cycles with data kept over 23 cycles, dating back to 1729. The corona has been seen to extend 40 solar radii from during the airborne observation of eclipses. The mass ejections exit the Sun as hot material and magnetic fields as self-contained structures. The geomagnetic storms that ensue when there is an interaction with the Earth in this way may disrupt power grids and satellite communications and it is a risk to manned spaceflight.

The Sun's own magnetosphere is termed the heliosphere. It is that region of space surrounding the Sun inflated by the solar wind over which the Sun casts a magnetic influence. It is quite elongated in shape as the Sun travels through space. Scientists reasoned that there must be a heliopause where the pressure of the solar wind balances that of the interstellar medium through which we collectively voyage. This has been observationally confirmed, literally in passing through part of it.

The two Pioneer and two Voyager spacecraft have crossed these divides and their ongoing flight paths are well worth a study on NASA's pages; these craft are the most far flung pieces of technology ever cast into the ocean of the cosmos. We are still in contact with both Voyagers, ongoing, since their launches in 1977. Voyager 1 is over 10 billion miles from Earth at the time of this writing. It is proudly called the Voyager Interstellar Mission now. In their active life among the neighbor planets they provided new views of Jupiter, Saturn, Uranus and the distant Neptune. In 1998, Voyager 1 became the most distant of human artifacts, outpacing the earlier Pioneer probes in speed and distance.

The Sun is actually moving at 12.5 miles a second toward the solar apex in the constellation Hercules and its planets are traveling with it. The speed conveys us at 4.09 AU a year in that direction and away from the solar antapex in Columba. The approximate positional points of direction are toward the star Vega and away from Zeta Canis Minoris. Determining these locations was another of Herschel's triumphs in 1783, justifying his celebration as the father of stellar astronomy.

On the grand scale, the Sun takes about 225 million years to make one revolution around the center of our local galaxy, moving at an estimated 140 miles a second. This has been termed the cosmic year. The delightful term "perigalaction" is the point in the Sun's orbit around the Milky Way when the Sun lies closest to the galactic center. This is also a feature of current events, the position the Sun presently holds. The direction is quite apparent with the glorious Sagittarian star fields revealing to even small telescopes the orientation to the center of the Milky Way.

Aurorae are the most visible and famous results of the Earth's magnetic field's interaction with the constant solar wind; that permanent radial flow of energetic and charged particles drags the solar magnetic field outward to form and maintain the interplanetary magnetic field. They have been seen since time immemorial and are part of folklore in the further northern and southern climes. One report from the 1st century CE describes the Roman emperor Tiberius dispatching a group of soldiers to battle a fire that turned out to be a blazing auroral sky. (Aurora is Latin for "dawn.")

Halley correctly suggested the cause to be a kind of "magnetic effluvia." Some sort of solar material arriving here and interacting with the Earth's magnetic field had been suspected since the early 1930s from ground-based observations.

A particularly strong solar storm with associated aurorae was described in early September 1859. They are caused by interaction between the Earth's magnetic field and charged particles from the Sun impacting the ionosphere whose layers surround the planet. They are distinct from mere airglow by their sporadic occurrence in the polar and sub polar regions, the color effects due to the emission of different gases. They occur for all planets with magnetospheres and, seen from above by satellite observation, form a glorious electric crown surrounding the magnetic poles of a world. Energy is strongly transferred from the solar wind to the magnetosphere during sub storms when the interplanetary and terrestrial magnetic fields merge at the dayside magnetopause in the course of a few hours. This is distinct from geomagnetic storms which are generally a more global disturbance of the Earth's magnetic field following violent activity on the Sun. They too are accompanied by sub storms and aurorae in reaction to solar flares and CMEs.

The heliosphere results from that region of space where the solar magnetic field is dominant and is inflated by the solar wind. Along the wave front where the pressure of the solar wind balances that of the interstellar medium lies the heliopause. This is visualized with that observationally direct evidence to lie beyond the orbits of even the most far-flung objects orbiting the Sun, on the boundaries of interstellar space.

The rotation of the Sun takes 24 days (yet famously an observed 27 days from our moving planetary observatory) and twists its huge magnetic fields to spiral form. The analogy is of a two-way rotating garden water sprinkler in constant shower on the lawn. In the deep past the Sun was hotter with a far busier solar wind that effectively exported much of the Sun's angular momentum and somehow broke its rotation to the comparatively sedate spin of this era. The vast proportion of angular momentum is stored in the orbital motion of the planets and comparatively little of it in the slowly spinning Sun. This is still not entirely expounded in scientific models for the formation of the planetary system. How so much angular momentum was placed in the planets' collective orbits with such a slowly rotating central Sun resulting is not fully clear.

It is still hypothesized that the solar system originally developed out of a primeval huge solar nebula whose material was possibly the result of a supernova explosion. Shock waves from such upheavals may have triggered the slow formation of a planetary system. Some extant members such as certain asteroids are relatively primeval and unchanged to this day.

Inside the Sun the relatively quiescent radiation zone extends to about 70% of the solar radius. There the temperature has dropped sufficiently to render the gas opaque. In transition, the innermost first tier of the convection zone gives rise, literally, to the super granulation cells, great convection cells up to 20,000 miles in diameter. Convection begins about 90% of the distance from the core to the surface where radiant energy passes through the radiation zone and is then converted into turbulent convection, ascending at a few miles a second in speed. The temperature at that point is down to approximately ½ million degrees K.

The face of the Sun takes on a bubbling and frothy appearance as the final outcome of that turbulent convection emerging as a photosphere. A granule is estimated to be about 1/300 the size of the super granule causing it from beneath.

It is active for about 8 minutes. In size they average about 600 miles wide or less and are separated by about 1000 miles on the solar surface.

The innermost central core is about a quarter of the volume of the Sun but packs 40% of the mass present. The core produces an estimated 90% of the energy that is eventually freed into space. The nuclear fusion roars away in the innermost core creating energetic particles and colossal radiation. At the outer part of the core the nuclear reactions slow and the temperature actually drops to 13 million degrees K. That difference, so high on the scale of temperatures, is the level required for fusion to proceed or pause. The core has been described as the perfect gas but with abnormal atoms. The density is twelve times that of solid lead but the conditions of pressure and temperature maintain it as a gas.

The pressure is greater than that of the Earth's atmosphere by a factor of 300 billion. It is also calculated that the time required for the energy transfer process, from core generation to free radiation into space, might take 170,000 years due to the random walk pattern of high-energy particles. High-energy radiation collides with the plasma in the radiative zone. About 25% of the way to the surface the energy transfer in the form of radiation beginning as gamma and X rays that, by collision, is red shifted to ultraviolet light. In visible light as one tiny part of the electromagnetic spectrum, the Sun is dark on the inside.

Beneath the chromosphere and chromospheric network there is a permanent pattern of sinuous chains extending over much of the solar disc, visible in Hydrogen–alpha light. It consists of small points brighter than the surrounding chromosphere. Magnetic field lines related to the super granulation cells lying beneath the photosphere form the network. The methods of flash spectroscopy reveal a very heterogeneous structure for the chromospheric transitional regions. From the visible surface to the corona there is a huge rise in temperature but major fall in density over a 10,000-mile zone of transition.

The solar corona is dominated by magnetic energy and that energy is stored in the coronal magnetic field. Over time it may be suddenly released by instability of lowered equilibrium in the overall field. Coronal mass ejections come from the more active regions that produce sunspots and frequent solar flares although they can derive from quieter regions that previously were highly active. During the solar minimum they are generally produced in the coronal streamer belt closest to the magnetic equator. With solar maximum they emerge from active areas over wider areas. Their ejection frequency is therefore related to the solar cycle: approximately one every other day at solar minimum and five or six a day at solar maximum. These figures are decisively lower limits as 50% of the changing face of the Sun is permanently turned away from the Earth by its own spin. There must be solar farside events, that is, events taking place on the side continuously turning away and turning into our sight.

CMEs generally possess a three-part structure consisting of a cavity of comparative low electron density, a dense core deep in the cavity and a bright leading edge. There are exceptions to this makeup. CMEs are not essentially connected to solar flares but the peak speeding up and the strongest radiation from the flare are often simultaneous. Flares are sudden releases of energy through a break in the Sun's chromosphere in the region of a sunspot.

These sunspots are merely lower temperature features that appear comparatively dark and transient. They are the easiest observed solar surface feature. In temperature they are some 1,500–2,000 degrees K less hot than the surrounding environment. In size they are generally between 1,500–30,000 miles in extent. They are also concentrations of magnetic flux and normally appear in pairs of opposite polarity. Sunspots inhibit the rise of convective heat and are a little lower placed in the photosphere. In appearance they are comprised of a dark central region termed the umbra surrounded by a lighter halo called the penumbra, with short, fine fibrils. The very largest groups can last for months and can reappear with the spin of the Sun bringing them back into view. They can be described as cooler plasma depressions, and this nature is revealed by their flattened and slightly lower appearance when viewed at the disc's extreme limb. That is called the Wilson effect. Sunspots also rise and fall in activity with an 11-year cycle or slightly longer. The underlying cause is thought to be a 22-year magnetic cycle.

Flares are sudden releases of energy through a break in the Sun's chromosphere in the region of a sunspot. These coronal holes are located at the poles of the Sun during solar minimum and open at any latitude at solar maximum. The so-called coronal holes have a density lower than the general corona by a factor of 100. Magnetic field lines emerging from the holes reach into space rather than loop back into the photosphere, forming the principal source of the solar wind. Flares may last from a few minutes to a few hours. Prominences are strands of relatively cool gas present in the corona that are seen as bright structures against the darkness of space but above the body of the Sun. They appear as filaments when viewed against the solar disc. Loop prominences, sprays and surges can occur with less hot material falling back from prominences as coronal rain. They can be suspended in the corona and average 8,000 degrees K. The active ones are associated with flares and sunspots.

The only time the solar corona is visible to the naked eye is in the glimpse during those precious minutes of totality with the climax of a solar eclipse. The corona comes into view as the Moon glides serenely into position. Baileys' beads and the diamond ring effect also accompany the beautiful spectacle of totality. Those glorious and dramatic moments formed mankind's sole notion of the corona's existence in the past.

Eclipses have this effect because the Sun is 400 times larger than the Moon and coincidentally 400 times farther away. Thus they both appear to be the same angular size, subtending half a degree on the celestial sphere. Including lunar eclipses, they are a statistical occurrence at least twice and at most seven times every year. There are an average of 237 eclipses in the course of a century, a quarter of them total solar in form.

One is recorded in an ancient Chinese account as far back as 2137 BCE and a Biblical reference in Amos 8:9, concerning the Sun going down at noon and the Earth darkening in the clear day, is interpreted to be another. A famous total solar eclipse in 585 BCE halted a battle between the Medes and the Lydians. This specific happening holds the distinction of being the very earliest historical event that can be tied to a precise time and place, to the minute and approximate square mile.

Flares are more common than the mass ejections. CMEs are more prodigious in energy and material but both export great quantities of particles away from the Sun and out into space. Giant solar prominences are more properly associated with CMEs. Intense flares do not usually accompany erupting prominences. At one time it was thought that flares drove these eruptions. They are more common but eruptions are much bigger in scale. Not unlike an escaping balloon, magnetic energy is released with the ejection. The attending erupting prominences and flares result from the energy of magnetic coupling of open field lines as they pinch below the rising prominence. Solar flares may actually be the result and not the cause of CMEs. Large and swift restructuring of magnetic fields in the lower corona are the prime causes of mass ejections.

Clearly, not all CMEs affect Earth or may be noticed at all. What is termed a halo CME is specifically a matter of perspective. It may be propagated in the direction of Earth and have an enhanced brightness because the material reflects sunlight in transit. It can appear larger than the Sun itself. An interacting CME persists on the scale of a few hours but may produce effects in Earth's magneto-sphere over the longer term. Such geomagnetic storms can result from either the faster CMEs interaction with the slower motion solar wind that has a steadier state. Their geoeffectiveness here depends on their speed and magnetic field strength. Such an ejection can compress the Earth's magnetosphere on the sun-ward dayside and extend to the night side tail.

Ten billion tons of plasma moving at a million miles an hour is not an exagger-ation for what the Sun can do The faster moving solar eruptions can overtake and proceed to devour their own companion structures in transit. A greater and more complex outward moving front of charged material results. These "cannibals" were first found by the SOHO (Solar and Heliospheric Observatory) mission in 2001. (Helios is Greek for Sun.) CMEs are the result of rapid large scale restruc-turing of magnetic fields in the lower corona. The energy is stored in stressed magnetic structures and following the event of a release in the field, reconfigure themselves into simpler arrangements.

In short, the Sun regularly hurls billion ton clouds of energetic, electrified and magnetized gas outward at speeds up to 800 miles a second. They are able to drive huge shock waves in their paths outward into interplanetary space. Their secondary effects and processes are more significant as "solar tsunamis." In the past there was no concept of this going on, beyond the majesty of the Northern or Southern Lights and St. Elmo's fire. It also reveals comet tails to be little more than interplanetary windsocks on a big scale. They have a small heated head of freezing hydrocarbons set in rock and ice showing a brief active transit when temporarily closer to the heat of the Sun.

From the mid 1960s until its resolution in 2002 there was an ongoing dis-cussion in astrophysics called the solar "neutrino problem." Whilst the Sun is reasonably well understood as a natural fusion reactor converting four hydrogen nuclei (protons) to helium with energy and neutrino output, there was a serious deficit in the quantities detected for the latter as predicted and required. This cast grave doubts that the inner workings of the Sun had been deciphered ac-curately. As it turned out, new information about travelling neutrinos changed the equation. It is now resolved that as neutrinos do possess mass they can and do change from the type that was expected to be produced in the solar interior into two other types that would not be caught by the detectors originally used. In the process of revision, the standard solar model was saved and a Nobel Prize for physics was earned.

Curiously, the incidence of genuine cosmic rays from the greater depths of the universe, i.e., galactic cosmic rays, are lower during a solar maximum as the strength of the ambient solar wind deflects more of those elementary particles from interacting with Earth. Equally, the flow of cosmic rays is higher in a solar minimum period with lower shock waves to deflect them. Whilst subatomic particles are accelerated by solar flares, the main quantities of energetic particles arriving here are mass ejections in source.

The solar corona can be regarded as a visible inner base for the solar wind. With the speeds and distances involved, a scaled down and dissipating CME will reach the distances equal to the orbit of Saturn in that 24 -day period in which the Sun coincidentally revolves on its axis. The complete structure of the corona extends permanently beyond the orbit of the Earth.

From our immediate viewpoint, less than 0.01% of the solar wind penetrates the magnetosphere, the region of space in which the Earth's magnetic field dominates that of the solar wind. It is distorted into a teardrop shape by the pressure of the solar wind on the dayside and on the night side draws out along a magneto-tail. This is a permanent but unseen companion to the Earth and may extend for over 10 times the distance of the Moon. The magnetic tail is the greater part of it. This is a decidedly dynamic structure, responsive to shifts in the solar wind itself and the orientation of the interplanetary magnetic field. All the planets possess magnetospheres to a greater or lesser extent. As may be expected, Jupiter has the largest and most active one after the Sun itself.

The Earth is protected in a magnetic cocoon that forms a cavity out of the solar wind that also brings the oppositely directed tail lobes into such close contact as to merge around the Earth. The disturbances in this magnetic connection event trigger sub storms lasting approximately one hour and occurring several times a day. Magnetic fields that are conducive to dayside merging can be brought about by CMEs. When the magnetosphere reconnects on the night side, energy is directed back toward the upper atmosphere and spectacular aurorae can result.

The inner Van Allen Belt is one of two doughnut-shaped rings of high energy charged particles forming radiation belts around the Earth. The closer one lies about 6,000 miles altitude or 1.5 Earth radii above the equator and the other about three times further out. They are both formed by high-energy charged particles trapped in the Earth's magnetic field. The magnetopause, the boundary between the magnetosphere and the magnetic field of the solar wind, forms at the distance where the solar wind dynamic pressure equals the magnetic pressure of the planet's field. It is 8–11 Earth radii upwind, depending on conditions.

The physics and kinematics of CMEs and much else about the Sun remains to be further analyzed. The Sun may even have had other stars that formed alongside it that could be identified through future research. Alpha Centauri is a similar type of star to the Sun yet this receives little comment in scientific texts.

Concerning the Sun, just a few features may be externally observed and calibrated; we can only rely on theoretical astrophysics for the inner structure.

The vital lesson is that it remains an ordinary star. Solar physics has much yet to learn. The plasmoid hypothesis entertained in this book as a thesis for the TCB's identity has no basis in hard evidence. It has never been put forward that the TCB was a special CME.

CHAPTER 7. ASTEROIDS AND THE ASTEROID HYPOTHESIS

The Tunguska explosion has technically been described as a superposition of a spherical blast wave by the terminal explosion of the TCB in the atmosphere and a conical ballistic wave axially symmetrical to the approach path. The pressure wave in front must have been colossal in scale. Its destructive impact on the taiga must have preceded the intense fireball and thermal waves. The temperature may briefly have been 5,000º Centigrade, comparable to the photosphere of the Sun or even higher for a brief time. Some speculations place it as high as 15,000º Centigrade or even higher. The wave(s) of heat were transient, the fires immediately extinguished by subsequent wave(s), although forest fires ensued at greater distances. The record is unclear how long the flames lasted.

Perhaps we have it there after all. Perhaps there was a low altitude airburst from a strangely heavy and isolated Beta Taurid meteor on which, as it dived rapidly into Earth's atmosphere, the pressure built to a level great enough to explosively and wholly vaporize the body in a rapid series of hot blasts. One lesser exploding meteor occurred over Snohomish, Washington State, in September 2009. It could be said that a much more minor but strictly similar event took place within 40 miles of where this is being written. In September 2002 the Vitim or Bodaybo event (named after another Siberian river and town) proved to be a bolide captured in passing by satellite observation. From the damage to the trees it was called a "mini Tunguska."

For our main event, a member of the Apollo asteroid group could be a possibility, as opposed to a radically unusual Beta Taurid meteor. The orbital positions of the Apollo group and their occasional proximity to Earth could make one of their member a candidate.

The enormous compression of air in front of a hypothetical asteroid entering the atmosphere and the near vacuum immediately following it represents a huge difference in pressure stressing the body. A strong drag force is exerted because the ram pressure in front of the body is so great. Shock waves are generated by the rapid compression of air in the path of the moving body. It is primarily this ram pressure rather than friction heating the air, which in turn flows around and heats the meteor. The TCB flattening out somewhat in shape at the finish may actually have been witnessed. This was merely one observer's words and must not be over emphasized. It is a choice to either give this great importance as a single crucial sighting or disregard it as an isolated and errant piece of testimony. A final violent dispersal into separate pieces may have been widely heard if not reliably seen. Witnesses agree there was a series of loud bangs but we cannot put them in a coherent sequence.

The hypothetical large asteroid speeding through the atmosphere would quite possibly "pancake" because a flat and thin shape is forced upon it by increasing air pressure. The thickening air of descent compressed the leading surface. Behind the entering body a temporary void was being created and swiftly adjusted by the flow of air. Here, the waves of concussion could transcend the term "sound" and issue an enormous rolling thunder. This was possibly the experience of the onlookers. The web pages of the Planetary Society strongly opt for an asteroid explanation and present a comprehensive description thereof. The size of the TCB is moved to a lower limit in their scenario, as is the mega tonnage of the blast. The damage to the forest is expounded not by a point explosion but by shockwaves impacting the ground. That the body did not strike *terra firma* but detonated in the air is completely agreed. They suggest that the fragmentation expected much earlier in the descent of a big asteroid was somehow delayed to the final moments, probably by its great tensile strength. By this reasoning, another few seconds and there would have been a ground impact. There must have been a major shockwave preceding the asteroid. In reference to computer simulations at the Sandia National Laboratories, a fiery wake would accompany the descent, something that concurs reasonably well with the collective eye witness accounts.

Their study in a news release in late 2007 oddly assumes that the forest at the time was not healthy and that the damage has been overestimated all along. I challenge that "winds were amplified above ridgelines where trees tended to be blown down" or that the forest at the time was in some sense unstable. This must be rejected because there has never been a suggestion that tree disease was prevalent at the time or that drought or any adverse growing conditions existed there. The thesis holds that the increasingly resistant atmosphere promoted an airburst that precipitated the downward flow of strongly heated gases. The value of 3–5 megatons is cited in their calculations.

It is not clear why there would be an interest in downplaying the effects of whatever took place at Tunguska. The topographic and ecological factors referred to did not create an illusion of forest devastation over 800 square miles of destruction. The trees were not weak and unstable or waiting to be felled by an explosion from the sky. The argument posits a small asteroid and suggests that the damage inflicted appeared exaggerated. Kulik would not agree from the earliest studies at the site and the pressure waves went round the whole world twice, clearly recording the scale of the explosion.

The fragmentation and breaking up of an asteroid is generally required to occur far earlier in its trajectory downward through the atmosphere. This is the consensus for the dynamics of hypothetically understood bodies. It remains a major problem in the whole modeling process of the Tunguska projectile. We could theorize a very dense, small metallic asteroid. We can make it spherical with a uniform density and set it spinning to distribute the thermal and pressure loads more equably — up to a critical point. Perhaps that will allow it to get close enough before totally self-immolating. Again, some fragments or a geochemical signature of the hot blast should be in evidence after all this study.

The force of the explosion and the temperatures engendered in the stricken forest beneath require more than a dense space rock of any mineralogical composition. It cannot pack and release that energy simply by burning up in the atmosphere and incidentally, leave no fragments or major dust residue below. There is no evidence of ablation as it soared towards us. The distribution ellipse of material and dust on the ground formed in other major meteorite falls is entirely absent. These features have been analyzed in other cases and we feel we have a good understanding.

Bolides interpreted as the airbursts from meteors occasionally occur on smaller scales and normally happen at those much greater heights. They are detected several times a year by military observation satellites. ("Bolis" is Greek for missile.) A satisfying description immediately presents itself here. We calculate that a stony meteoroid about 30 feet in diameter could produce an explosion of about 20 kilotons under guesswork of the other accompanying factors.

The individuality of the body seems at odds with efforts to link it up to a known (or for that matter a sporadic) meteor stream. The TCB is hardly a member of a group and bears no associations we can place. There was no accompanying storm of cosmic material that had wandered the solar system for eons, a large piece plunging dramatically Earthward one day as the world turned its face to the Sun along the longitude of a Siberian dawn.

Proposing that it was interstellar in origin and not a locally grown product adds little to the discussion. The odds of it surviving, remaining potent and arriving here are vanishingly remote. A denser, rockier object is suggested by the proximity to the ground of the terminal blasts that sounded "like artillery fire" or even "falling stones." We have encountered a truly unique natural object. It does not bear scrutiny for an unusual meteor or a piece of a comet. There are too many objections to that. To be fair to all possibilities, a unique but unnatural object may have shattered the Tunguskan morning. There is the whole alternative to the elusive natural body we cannot define, an artificial, contrived and unnatural body whose theme we will explore.

One up-to-date college textbook entitled *Explorations: An Introduction to Astronomy* has a chapter on giant impacts that definitively states that at Tunguska, an asteroid exploded in the air, leaving no crater but curiously left trees standing at the center of the damaged area. That the event caused two human deaths is now mentioned. A description assures the student, referring to computer simulations, that an asteroid could and did behave this way, violently exploding yet leaving no crater or ground fragments in this most recent of giant impacts. A cometary origin is disparaged. Attributing the bang to the detonation of icy material following the passage of the dazzling blue bolide has not proved viable.

That some other hugely diffuse, dim and unknown comet was responsible with a very compact nucleus constituting the TCB is a useful speculation. It was non luminous or certainly low magnitude out in space. How could it have escaped notice from vigilant astronomers? There was no direct warning of its approach.

Allowing for a small high-speed specimen that is predominantly extinct or cool, one could propose an inert comet with a dark mantle fully approaching in daylight. By the observers' line of sight it came out of the Sun in direction. This returns us to the dynamics of atmospheric entry. Whatever it was should have heated, vaporized or broken up at a much greater altitude, even if equipped with a cold, thick stony outer envelope for initial protection as it started downward. It very nearly hit the forest. The ram pressure must have been immense considering the strong seismic waves caused by a body that unequivocally did not strike the ground. This was the first time a celestial body had been detected seismically and successfully by the instrumentation existing in 1908. How could it have got downward from space before pressure, friction and mounting temperatures somehow destroyed it so completely?

The physics of atmospheric passage is different for the massive bodies than for common tiny meteors because aerodynamics works differently on large bodies. One often gets to see a flash of a meteor during a few hours beneath a clear night sky. Quite simply, a larger body is calculated to mostly survive the passage, possibly fragmenting in transit and impacting the ground, given that it has the cohesion to reach that close which is the case.

Those records of noctilucent clouds, solar haloes and multicolored twilights seen in Middle Asia, European Russia and Western Europe in the days leading up to 30 June and coming before the fall could prove significant. It is another mystery apart from the inherent nature of the TCB on which we concentrate. No specific object was seen coming or in any sense anticipated, according to subsequent enquiries at numerous observatories. We may speculate that some imposing but tenuous body of material was already encroaching on the upper atmosphere and having noticeable effects prior to the main event. That had been suggested in explanation of the illuminated twilights after the event by a scientist unaware of the drama taking place.

Surely there were active observatories that should have seen anything substantial coming or detected its occulting effects on background stars if it was of a dark nature? Perhaps it was altogether too small and non-luminous in the night sky or entirely came out of the Sun's direction. Possibly we are dealing with energy or magnetic fields rather than a material body approaching. We will entertain that deduction.

Some eyewitness accounts indicate something larger than the Sun and perhaps briefly brighter or "a moving star with a fiery tail." We safely assume that the TCB only took on this brilliance traversing downward in the thickening atmosphere of Earth. It was presumably dim, small, and successfully escaped any

notice in approach. May we conclude that it took a full and speedy daytime approach to Earth, practically out of the Sun in direction? We have testimony of
a "fiery body like a beam" and that "the bright body seemed to smudge and then
turned into a giant billow of black smoke" at the finish. Estimates of its inherent diameter vary widely, but smaller has been the consensus. According to one
account it appears to have been seen from Vanavara at about 12 miles high and 1
mile wide in a close and fiery passage.

Did some stream of superheated gas encapsulate and protect a tiny dense nucleus that blew up so powerfully that it left no surviving pieces? Our old friend
Kulik had suggested something similar to the first clause.

Some form of cometary theory tends to dominate. If we require one with a
deuterium component to explain the powerful temperatures and possible nuclear nature of the blasts, then it decisively does not fit our normal model for a
comet's structure. The ice and rock conglomerate of the typical example is well
understood nowadays and an unknown explosive chemical compound including
deuterium is unlikely in the extreme.

In the aftermath of the event the observatory at Heidelberg had several celestial photographs severely clouded by the bright night sky following 30 June.
The bright twilights were seen as far west as Spain and Bordeaux, France. Dusty
interference in the atmosphere may have kept up until mid August 1908.

With the asteroid theory it is equally difficult to account for the complete
lack of the tiniest fragments from the parent body. How could a major bolide of
any given stony, carbonaceous or metallic composition detonate aloft with such
heat that it leaves no obvious vestiges ever to be found? Only the advanced techniques of chemical analysis and electron microscopy present us with any fragmentation that could be held as components of the vanished TCB. If the asteroid
were large enough to survive most of the way through the thickening atmosphere,
it simply should have partly made it to the ground and struck like the Barringer
impact. There the conversion of energy into heat mostly vaporized the incoming
asteroid and created a major crater. A surrounding debris field of nickel-iron fragments was left over in the Arizona impact. Also, heed a warning when meteorite
prospecting: molten terrestrial material splashed from a meteorite impact can
cool and solidify in the area. These tektites may easily be mistaken for meteorites.

In the case of Tunguska we have instead a terminal cataclysm aloft. Some
pieces of a greater or lesser size should survive a downward fall to the ground
from such a low height of explosive fragmentation. The TCB somehow engulfed

itself in an instant. There was no breaking up along the way. This mystery applies to any normal speculation for the composition, mass and density our hypothetical asteroid requires. Others disagree. It has been reasoned and calculated as fully consistent with the explosive disruption of a 60–100 meter asteroid of the common stony class under the extreme stress and pressure conditions present (Chyba et al., 1993). This is controversial with the conditions of no tiny fragments or dust surviving to the ground beneath. The model hinges on the forces opposing the descent of the body and ram pressure overcoming the body's forces of cohesion. Following the buildup there comes a momentary explosion with a full release of energy.

In the early 1980s, Zdenek Sekanina derived an asteroidal body 100–200 yards in diameter entering the picture at a low 5 degrees to the horizon and an azimuth of 110 degrees and proceeding at 67,000 miles an hour or approximately 18 miles a second. This is coincidentally equal to the average speed of the Earth's path around the Sun. His paper strongly rejected the comet hypothesis and drew heavily on the eye witness accounts. It favored an Apollo group asteroid for firm identity.

It is not clear that we can ever know the speed or direction from space it took in getting here. The data is non-existent and no amount of review is going to reveal facts never recorded. Statistically, the likelihood of an asteroid passing close to the Earth's orbit around the Sun and posing serious danger of an impact to us is far higher than that of a comet. Whilst comets may be extremely common overall, out in the deepfreeze of the Oort Cloud, the solar system is populated by a nearly a quarter million known asteroids. They mostly orbit in the loosely named asteroid belt lying between the orbits of Mars and Jupiter and are strongly influenced by the latter's gravitational field. Jupiter's gravity is second only to the Sun in strength and influence as is its magnetic field.

There was a graphic proof recently that "Jupiter grazers" exist. There are more asteroids than comets in the inner solar system on a regular basis despite the profusion of the former in the bigger picture. (We define the inner solar system as anywhere within the orbital distances of Mars.) Chyba and Zahnlev (of NASA and Ames) calculate that a carbonaceous asteroid would have exploded at just under 9 miles altitude and a stony body around 5½ miles up, which is consistent with the case at hand. These stony asteroids are also the commonest in type overall. An iron based projectile would, by their estimates, mostly if not entirely reach the ground.

Kulik always maintained this and searched determinedly for a massive iron body or its myriad parts. Meteorites and craters do not apply to Tunguska studies, no matter how hard we look or maintain that such trace entities must exist somewhere. An asteroid may be the more likely candidate by probability, but we cannot rationally speculate on its previous orbit or planetary path. An Apollo or conceivably an Aten class asteroid are speculative possibilities. One is tempted to call it a very strange Beta Taurid and hang it all up. Say the small companions all vaporized high in the sunlit sky like good little accompanying Beta Taurids and the problem of Tunguska is resolved — although there is some major inexplicable "dust" left in the air.

The atmosphere has its limits as a shield to both space rocks and harmful radiation and the entry angle and speed are among the major considerations for any modeling process. Somehow it let this one through almost entirely intact. Can we assume that the TCB was homogenous and relatively high density in nature, like some uniformly iron asteroid? The early researchers subscribed to these hypotheses. Alternatively, did the crust of dark, cold cometary core ablate away swiftly with a high-speed entry into the atmosphere? It then decelerated whilst heating up and abruptly exploded with tremendous force. Then Earth was impacted by a piece of a comet with a strange energetic nucleus and the case is again closed. Academically, the astronomers can debate the Beta Taurid or Comet Encke character for it. They can be construed as the same things over time from our knowledge of the Taurid Complex.

What are asteroids, exactly? Asteroids or minor planets range much further afield than the inner solar system and more generally roam out there in the first place. The main belt is defined as asteroids with semi-major axes of 2.2–3.3 AU, taking 3.3 to 6 years to orbit. Mars and Jupiter have positions averaging 1.52 and 5.20 AU from the Sun. They are colloquially between Mars and Jupiter in position.

A projected 75% of all asteroids occupy the main belt and there are 20–30 reliably recognized families and scores of less certain groupings. An estimated 33%–35% of the main belt bodies are such family members. Asteroids are best described in terms of their orbital elements, there being several methods of ranking them. The Apollo group, for example, are defined as having their mean distances from the Sun equal to or greater than 1 AU and a perihelion distance less than Earth's aphelion distance. They are one of three near Earth asteroids groups whose notable members include 1566 Icarus and 1995 CR that pass closer to the Sun than Mercury's orbit. The Aten group is defined in having semi-major axes of orbit less than 1 AU. The Amor group has semi-major axes that lie between the orbits of Earth and Mars, and the two tiny moons thereof may be captured Amor asteroids. 433 Eros, with which we have the closest familiarity, having landed a vehicle on it, is also an Amor group member.

These minor planets are of three broadly distinct compositions or a mix thereof and are distributed at varying distances from the Sun. Stony S-types are commonest in the inner parts of the Belt. The carbonaceous C-type predominate in quantity and the third metallic M-type are least numerous overall. M-types are approximately 10% of the sum total and are generally concentrated in the middle of the belt. We reason that the metal type asteroids are the leftover of the central core zone of shattered small bodies in the distant past. The mantle and crust of the vanished and disrupted small planet(s) possibly gave rise to the C and S –types.

The very largest asteroids are spheroid in shape and probably differentiated in structure, i.e., they likely maintain a mantle and crust with a very tiny and negligibly quiescent metallic core. Small pieces of 4 Vesta have made it to the Earth's surface and are available for direct study. This is the only known and recognized minor planet of which we possess a specimen. Vesta is unusual for its strongly basaltic surface and obvious volcanism in the past as is 1459 Magnya. NASA's Dawn mission is scheduled to encounter Vesta and Ceres in August 2011 and February 2015 if all goes according to plan. The NEAR spacecraft was successfully landed on the asteroid 433 Eros in 2000. The Deep Impact mission fired a probe directly into Comet Tempel 1 in July 2004. Some very exciting pictures and informative data were returned. Asteroids can also be the parents of meteors, ultimately deriving along with everything else in the solar system from the primeval solar nebula. Was the TCB some such piece of primordial material?

Piazzi discovered the first asteroid (meaning "starlike," from the Greek "asteroeides") in 1801. Herschel first applied the term. This original find proved to be the largest of all asteroids and was named Ceres by its discoverer. Later it was designated 1 Ceres and was found to be just under 500 miles in diameter. Only Vesta is narrowly a naked eye celestial object around its opposition. They were as a class of object quite unknown to the ancient or medieval worldview and are entirely telescopic with the qualified exception of Vesta at opposition.

Four were discovered in the first decade of the 19th century and the matter rested there for forty years until the fifth discovery in 1845 by a determined Hencke. He named it Astrea. A collective 300 were known by 1890, the same year that Max Wolf introduced astrophotography to this branch of astronomy. He was rewarded with a comet recovery and locating asteroid 323 in 1891 that was named for him. His pupil Reinmuth only exceeded his record of 248 discoveries by 1932. The 1,000th was seen in 1923. There have been new discoveries every year since the mid 18th century with the understandable exception of 1945. At times the trails of asteroids on photographic plates even achieved a kind of nuisance value for astronomers trying to obtain better images of deep sky objects.

Olbers, the discoverer of Vesta in 1807, suggested that the varying apparent magnitudes of Ceres and Vesta were due to irregular rather than round shapes. The discovery of the asteroids is one of the most enterprising tales in the advance of planetary science in Europe engaging some very talented intellectuals and state of the art telescopes of the era. The astronomical association of observers proposing some systematic searches of the night sky over Germany during the early 19th century called themselves the "celestial police." Gauss developed the mathematical technique for calculating orbits that confirmed the true nature of the motion of Ceres a year after Piazzi's first sighting. He produced some formidable work in mathematics and magnetic field studies.

Kepler had suggested that on principle there could be a body between Mars and Jupiter. He had concluded that the spacing between the orbits increase in distance from the Sun and was working prodigiously on the long term observational records, principally of Mars that he had received from Tycho's efforts. Commenting on a conspicuously large gap beyond the fourth and fifth worlds, he wrote in 1596, "Inter Jovem et Martem interposui planetam" — "between Mars and Jupiter I would put a planet."

The Titius-Bode Law arithmetically describes (with a few glitches like the mean distance of the planet Neptune) the average distances between the planets. It was a first elucidated in 1766. The range of distances suggests a position that could be occupied by world(s) unknown. The separation of 2.8 AU from the Sun would prove to be positionally descriptive for the main asteroid belt.

In terms of mass the whole assembly of asteroids would not collectively equal that of the Moon. It is thought unlikely now that the asteroids derive from a single previous and lost world. Such an imagined body was hypothetically placed between the paths of Mars and Jupiter in the past by a boldly original hypothesis of the noted Olbers. Later is was given the retrospective name "Phaeton." That body would have been notably smaller than any inner planet or major satellite in mass. The real asteroid 3200 Phaethon was the first discovered by spacecraft

(IRAS 1983) and is at the same time an Apollo object, extinct comet and source of the mid December Geminid meteor shower. Note the spellings and that the hypothetical Phaeton probably never existed.

When the Galileo spacecraft en route to Jupiter first imaged the asteroid 951 Gaspra in 1991, it was both a revelation and relief to scientists to finally have a decent picture of the real thing. It did look as expected and incidentally resembled an 11-mile potato in appearance. Gaspra is an S-type minor planet and one member of the Flora group. Based on the comparatively low quantities and small sizes of craters on Gaspra it is thought to be relatively very young. It has not been exposed for longer than 200 million years. The same spacecraft photographed 243 Ida two years later and discovered a small companion of its own, subsequently named Dactyl. Binary asteroids were revealed to occur, a condition not generally predicted due to their low individual masses. Ida is thought to be a primitive body and both its density and appearance are consistent with S-type asteroids in general. Not only are the two diminutive natural satellites of Mars captured asteroids but numerous smaller moons of the gas giant planets strongly suggest themselves to be of this origin.

The Uranian moon Miranda may have been broken down and reformed on an estimated five occasions. The Saturnian satellite Phoebe was thought until recently to be an expended cometary nucleus and certainly a primitive body. It is furthest out of Saturn's retinue with a retrograde orbit and engenders the largest and outermost tenuous ring, tipped 25° to the plane of the inner ring system. It may be responsible for dumping material on the other Saturnian moon Japetus, explaining the darkness of that satellite's leading edge. In size it is 135 miles across and may be along captured KBO. The Spitzer Space telescope provided this information in late 2009.

The two Trojan groups of asteroids permanently maintain an equal distance from the Sun and Jupiter in a gravitational resonance and hence, equilateral triangle. (See Lagrangian and the five Libration Points.) In 1990, Mars was found to have Trojan asteroids of it own, in their own permanent orbital resonances. Asteroids vary widely in albedo, their inherent reflectivity, but tend to the lower values as very dark rock.

The NEAR mission (Near Earth Asteroid Rendezvous) went on successfully to plot the mass and density of 253 Mathilde by measuring the gravitational effects on the space vehicle from a close point of 750 miles and take more beautiful images. The surprisingly low density for Mathilde leads to the conclusion that it is a loosely bound collection of smaller rocks in nature. In February 2001, after its immense journey, NEAR successfully approached and finally touched down on 433 Eros, a remarkable technical achievement. The NEAR probe, its work done, has taken up permanent residence on Eros.

In the outer solar system, the largest known asteroid is 2060 Chiron with an approximated diameter over 100 miles. The perihelion is just within the orbit of Saturn with a far point to almost the distance of Uranus, never venturing very sunward on its journey. Clearly, many more planetoids and inwardly wandering Kuiper Belt Objects await discovery in the deeper reaches. Following Chiron's discovery in 1977 and noting that it had been recorded several times and as far back as the late 19th century but unrecognized, it began to show distinctly cometary characteristics in 1988. Over the next few years it showed both a coma and a tail. At the distance of 12 AU from the Sun it brightened by 75%. It is the prototype of a Centaur, partly asteroidal and partly cometary in nature. A few others have been identified with eccentric, long term and probably unstable orbits. They may either be ejected from the solar system or more likely with Chiron, eventually become a short period comet. It was originally a refugee from the Kuiper Belt and also bears the comet designation 95P/Chiron. (The name should not be confused with Charon, the satellite of the dwarf planet Pluto.) The scientific lesson here is that there is no strict demarcation between asteroids and comets in some rare cases. Closer to home, there are 3 known non-Centaur asteroids classified as main belt comets by their orbits and permanent position in the asteroid main belt.

5145 Pholus is poetically named as another Centaur and mythologically the brother of Chiron but has not shown cometary behavior. According to the The Minor Planet Center and consistent with grouping asteroids by orbits, Centaurs are defined as having perihelia beyond Jupiter and a semi-major axis less than that of Neptune. Its path reaches beyond Neptune and has the reddest surface of any known planetary body, earning the nickname "Big Red." Spectroscopically, this is attributed to organic compounds on the surface. Also regarded as an inwardly encroaching KBO, it is estimated that its 92-year orbit has not taken it within 1 AU of a planetary body since 764 BCE and will not do so again until 5290 AD. Both Chiron and Pholus are considerably larger than any cometary nucleus.

The clear and sometimes alarming sightings of comets appear in many ancient records including lucid descriptions and illustrations.. They can lay on far more spectacular shows and cause grave alarm to the populace. They have been historically associated with wars, pestilence and disaster. As noted, these included the death of Caesar in 44 BCE and the Norman invasion of Britain of 1066 AD. A Babylonian tablet from 1140 BCE and a Syrian dated 146 BCE describe comets. The destruction of the Minoan civilization on the island of Crete about 1475 BCE was more likely a tsunami resulting from earthquake and volcanic eruption than of celestial cause.

Asteroids may therefore be seen as the rocky debris from the early development of small members in the inner zones of the planetary system, many as failed plantesimals. Previously they may have been more uniformly distributed across the solar system than the broad belt extant today. Many were probably entirely ejected gravitationally form the system by the "Jupiter effect." Formally there was a competing theory of fragmentation for their origin, which visualized the destructive break up of a small world and its debris still in orbit in many pieces. Today the theory of accretion is far more in favor as an explanation for their original formation. This holds that asteroids were more likely formed by the failure of small bodies to develop into a major rocky planet from the swirling cloud of gas and dust of primeval planetary formation.

Another engrossing theory describes an overall outward migration of the orbits of the planets since very early times. Asteroids, by mass, only play a minor role. The primitive natures of some asteroids testify to their very great age. The dominance and disruption of the Jupiter gravitational field played a major part and continues to do so. Accretion is generally accepted as a prime hypothesis now though they have clearly undergone major impacts since. The gentler accretionary collisions in the deep past gave way to catastrophic disruptions that followed on. Both theories are valuable but fragmentation came later on the originally accreted bodies. No solar system object should be seen in complete isolation for its formation.

It is not impossible to think that there may once have been a smaller rocky planet beyond Mars, potentially of comparative size. The asteroid belt itself has Kirkwood gaps, dips in the distribution of main belt members due to orbital resonances with Jupiter, first discovered in 1857. The physical size of Mars itself may have been limited in development by the disruptive Jupiter gravity. At the time of this writing there are twenty known minor planets with retrograde orbits making a distinct minority group and again, they are of uncertain origin.

It is fundamentally thought that the iron meteorites derive from the core, the stony-irons from the mantles and the achondrites from the prior crusts of major bodies that originally formed by accretion. Some larger bodies may have become heated sufficiently for their interiors to melt and geochemical processes to commence. Sources for that heat may have included the radiogenic decay that still drives the plate tectonics here on Earth. For example, metallic iron separated out and sank to the center, thus forming an iron core and the lower density basaltic lavas were forced to the surface. Subsequent disruption wholly or partly ripped away their crusts, exposing the iron cores. There are probably both original condensations from the primeval solar nebula and more collisional fragments forming families of asteroids by type of orbit, e.g., the Hirayama group.

On a clear night you can see the flashes of tiny meteors zipping momentarily across the firmament and burning up. They average once every twenty minutes and can be sporadic, i.e., not belonging to any recognized group. Actual showers from well defined radiants are familiar and reliable enough to earn names like the Leonids whose source is the comet 55 P/Tempel-Tuttle or that most reliable group the Perseids, spawned by Comet Swift-Tuttle, the largest solar system body that regularly approaches Earth. It will pass within 0.015 AU in 4479 AD.

The names of the meteor radiants are formed from the apparent place of constellation origin or individual star by pure line of sight. The radiant point is indeed an effect of perspective like viewing the railroad tracks from a bridge above the paths of the trains. They can produce delightful shows of celestial pyrotechnics under the right conditions of clear preferably moonless evenings on a schedule lasting several nights. August happens to be the best month to see them.

Basically, there are known patches of dust and rock that lie in the Earth's path and are intercepted at regular and fully predictable annual periods. We continually encounter this debris that originates from active and decayed comets. Some small parts of Comet Halley are visible every year in May and October. The stream is a complete loop of material derived from P/Halley, regularly visiting us as the Eta Aquarid and Orionid meteor streams.

The Jewish Talmud of the 1st century AD refers to a "star which appears once every seventy years that makes the captains of the ships err." We may assume from that periodicity that this identifies Halley's Comet, two thousand years ago. The Rabbinical writer was known to have been active during the comet's apparition of 66 CE.

The Black Stone of the Ku'bah in the Great Mosque at Mecca is believed to have descended from Heaven and this is figuratively accurate. Japanese and Indian cultures also have meteorites revered as holy relics.

One of the many unknown factors for the Tunguska projectile was its tensile strength. We perceive asteroids as flying mountains of dense, indestructible solid rock or raw iron threatening Earth with wholesale destruction. Certainly, the larger ones are, taken at face value. Comets, conversely, can be viewed as looser icebergs with a rocky nucleus springing to activity when approaching and becoming warmed by the Sun.

When the NEAR mission flew by asteroid 253 Mathilde in June 1997 it was revealed to consist more of a loosely packed rubble pile. Approximately 50% of its interior volume is empty space. With dimensions of 36 X 29 miles it was also found to have a very low albedo, a mere 3% reflection of the light striking it from the Sun making it a truly dark body. Our close up experience is limited but jointly with the nucleus of Comet Borrelly, this is the darkest body ever encountered. The slow rate of spin for Mathilde is far less than the 2–24 hours typical of asteroids. Imagery of 60% of its surface in the course of the mission's passage was possible. Its measured density is less than half that of a normal carbonaceous chrondite and the interior is assumed to be quite homogenous. It must have some structural strength. The low interior density has acted as a poor transmitter of impact shock preserving the surface impact features to a very high degree. The angular edges of the craters strongly suggest that major chunks have been spalled away in the past. The surface is mostly phyllosilicate materials.

Much was gleaned about Mathilde from its gravitational interaction with the passing NEAR probe as the target body journeyed along at 11 miles a second in a 4.3-year orbit, providing good hard information and spectacular imagery of an asteroid up close for the third time.

Cometary dust has been collected as close to home as the upper atmosphere. Also, a tiny sample of Comet Wild 2 was also successfully delivered on board the returned probe from the Stardust spacecraft. The ongoing mission is due to flyby Comet Temple 1 for a second look at the changes on the surface wrought by the impacting probe.

The falling of cosmic dust is not so rare and remote either. The Earth gains at least twenty tons a day in weight from regularly falling meteoric dust. The ancient surface of the Moon is positively coated with it, emphasizing the great and undisturbed age of the lunar surface.

There was no obvious tangible remnant of the TCB itself for the early researchers, only the physical and chemical results of powerful shock and heat waves.

Tree resins from the year 1908 have shown micro sized particles of 14 distinct elements presumably trapped at the moment of the event. The scanning electron microscope revealed abundances of iron, calcium, aluminum, silicon, gold, copper, titanium and nickel showing evidence of heating and melting. The high proportion of nickel relative to iron in the tree resins resembles that of a meteor's chemical signature. There are some rare earth concentrates in the peat layers that were being laid down at the time. The magnetite and fused silicate globules in the soil samples show concentrations higher for 1908 than in layers before or after. The tree rings examined for that year show a rise in carbon-14 but that may be attributed to the solar cycle. The tree rings are readily available for fresh study to this day, and they are not affected in a manner that suggests a nuclear explosion as cause. The temperature was not so high as to be compared to an atomic blast. The forest was damaged but not annihilated by temperatures of that spectacular scale. There is evidence of accelerated growth following summer 1908 and a higher proportion of iridium in the bog layers of the immediate area were laid down in the season, conceivably of extraterrestrial origin. Magnetite may form on the exterior of a meteorite as ablation occurs in an oxygen rich atmosphere and the glassy spheres of silicate minerals could be typical of stony meteorites. There remains a notable lack of any such products of ablation anywhere along its possible path. Some samples located at Tunguska were composites of both magnetite and silicates, strongly indicating that they were formed together. It was never concluded that any of this material was part of the TCB.

Iridium is relatively rare in the Earth's crust but is common in meteorites. An Antarctic ice core sample taken in 1974 showed that at the depth corresponding to 1908 and a few subsequent years there was four times the normal concentration there. The microscopic examination of Tungus soil samples revealed tiny spheres of meteoric dust mostly composed of magnetite, an iron oxide mineral possibly formed by ablation in an oxygen rich atmospheric passage. Other spheres were glassy, indicative of silicate material such as that present in stony meteorites, and some spheres were composites of both types suggesting a simultaneous formation. With a major icy component, this tends to lend credibility to the cometary hypothesis.

In Omsk during the 1980s, a chemist named Dozmorov analyzed soil samples from Ostraya Mountain and found that the ratios of rare earth elements were very unusual especially with the comparative large quantity of ytterbium. He broadly suggested that a most unusual chemical composition and a non-natural substance had formed the TCB. Further work by ITEG scientists scrutinized micro samples of peat by scanning electron microscopes. Metallic particles, titanium, rhodium and aluminum with some copper and manganese suggestive of artificial construction, even a little gold was found (Agafonov, Institute of Geology).

The major iridium layer at the geological K/T boundary is one of the signature pieces of evidence for the impact resulting in the demise of the dinosaur. It was the violent change in environment triggered by a major asteroid blow or possibly several simultaneous events that caused that mass extinction in Earth history. The buried Chicxulub crater itself is about 110 miles in diameter, originally derived from a body at least 6 miles in size. Among the things indicating an impact origin are shocked quartz and tektites and the gravity anomaly of hugely compressed rock in the region. It remains one of the largest confirmed impact structures in the world and was a major discovery by the geophysicist Glen Penfield. There are also iridium traces at the South Pole corresponding to the period immediately following 1908, but it is not clear how the accumulation occurred there.

It is not out of the question that some deeply buried pieces of the TCB are lodged beneath the Great Hollow or that Kulik's hands-on excavations were fundamentally on the right track of investigation. Major efforts of georadar techniques might be usefully employed in the future. On Stoykovich Mountain there lies a strangely large ten-ton stone near the epicenter known as the Deer Stone, originally found by the Anfinogenov party in 1972. This was briefly interpreted as a piece of the TCB but that has now been completely ruled out. It is terrestrial in origin. Quartz samples from its surface revealed by their thermoluminescence that the stone had been exposed to hard radiation which can only have been the thermal flash of our elusive object from space.

Meanwhile, back in Siberia, could there have been a component to the explosion like a subterranean methane eruption under pressure that the fireball merely set off?

I have read up on this at some length. A ground-up explosion alone in some rare geothermal gaseous detonation is obviously inconsistent with the observed descent from the sky integral to the Tungus event. The thousands of smashed and charred trees were not uprooted or blasted from beneath in any part of the affected zone. More distant ones clearly had their tops blown away and their lower parts left comparatively whole. The explosion from the sky ignited and charred the immediate forest but the ensuing blast wave put the fire out very shortly afterward in the innermost regions although the exact order and paces of those events are undetermined. We will shortly consider alternatives to any comet or meteor theories.

It would be engaging if Lake Cheko lying five miles NNW of the epicenter with an axis aligned toward it were shown to have formed in the cataclysm. Maps from that era are too imprecise to reveal if that lake was in existence prior to 1908. The lakebed has a conical shape consistent with an impact crater. Its dimensions are some 500 yards wide and 50 yards deep but preliminary investigations suggest a far greater geologically age. Even if it were conclusively shown to be of impact origin it could have been formed millennia ago by the same process. There are trees surrounding it older than a mere century.

There is a third-hand oral account that attributes the lake's formation to the Tunguska explosion, but that derives from a supposed remark by the Evenki named Aksenov, who had an interest in the area at the time when the shamans were declaring the fall zone sacred. Florenskiy was convinced as far back as 1963 that this was not the case. In his own words, the depth of the silt layer on the lakebed indicates an ancient origin for the lake, tentatively estimated at 5,000–10,000 years. There is the evaluation from other studies concurring that the silt deposits date from far earlier. The lake may be impact in origin and geological atypical in the region, but it is far older than a mere century.

It remains possible that some small extant piece of the body lies at the bottom of the lake as suggested by acoustic investigation and a magnetic anomaly along the lakebed, but this is quite speculative. The University of Bologna has done some research on this question.

There is one case of a fully documented and better understood fall to Earth and that is the Sikhote-Alin meteorite, also in Siberia. It landed at 10:38 a.m. on 12 February 1947 near Vladivostok and is far more straightforward as a study. The name Sikhote-Alin derives from the Siberian Maritime Province and mountain range where it occurred.

There are some 200 craterlets, the largest being 85 feet in diameter and 20 feet deep. Not only were they numerous but previously record amounts of meteorite material were recovered over a period from a prime iron body of an estimated original 70 tons in weight. Pieces of it are available for sale online. Collectively, 23 tons of material have been recovered with an estimated 47 tons still lying there. The researcher Krinov calculated for a single body shattering at a height of 3.6 miles with the largest recovered specimen weighing in at 2 tons. It is now on display in Moscow. A 30 pounder was discovered lodged in a tree.

Technically, the body proved to be structurally a coarse octahedrite. Chemically, it is a Group IIB of 5.9% nickel, 0.42% cobalt, 0.46% phosphorus, about 0.28% sulfur and other trace elements and the remaining 93% iron by composition. This makeup was a safe assumption because so much material had survived the entire passage through the Earth's natural protection of gases and reach the ground. There was plenty of meteoric dust present in the strewn field too, quite unlike Tunguska where there was no such. Local bystanders are quoted as saying they thought it was a nuclear attack.

There is also the curious recording from the South Pole where Shakleton's Antarctic expedition witnessed an extraordinary bright aurora about seven hours before the blast of 30 June 1908. The team was located near the Erubus volcano at the time. Clearly, the aftermath of the Tungus Event impacted the ionosphere and may have disturbed the Earth's magnetic field locally (?) for a few hours and its climate very slightly for a few years. A radio operator out at sea suffered minor burns with the electrical overloading of circuits of his equipment. The evidence of major magnetic interference is clear. The graphs of magnetic activity found at the Irkutsk Magnetic and Meteorological Observatory in 1959 reveal a major magnetic storm immediately after the event that lasted four hours.

As for a scientific embarrassment, soil samples from the ground for the Sikhote-Alin event were once accidentally confused with Kulik's Tunguska soil samples that had been in storage. An erroneous attribution to an iron meteorite for the TCB by the microscopic identification of tiny globules of magnetite with traces of nickel and cobalt was made in 1957. This conclusion applies to Sikhote-Alin studies only.

Kulik had many serious logistic difficulties involving everything from the terrain to the mosquitoes, likened to flying alligators, and guides who refused to go closer to ground zero in superstitious dread of the disgruntled thunder god's home turf. Temperatures drop to 50º F below zero in the colder months when the land is not a treacherous bog. As mineralogist, scientist and meteorite expert, our man in Tunguska did a huge and complex job of fact finding, interviewing, photographing, surveying, marking and traversing the territory with severely limited means. His work was methodical and is still helpful to future investigators.

Kulik himself joined a volunteer military regiment and died in the Second World War. He had been a revolutionary, soldier, teacher, scientist and explorer in his life. A street in Vanavara was named after him and a Soviet postage stamp commemorates his achievements. Both the minor asteroid 2794 Kulik and a crater on the Moon's farside were subsequently named in his honor. He was not above reproach or professional criticism but his scientific competence and sheer determination saved the great enigma of Tunguska from sinking into obscurity. The Academy placed a simple memorial on his grave at Spas-Demensk in 1960.

In the immediate post war period and after the death of Vernadsky in 1945, Fesenkov became Chairman of the Academy's Committee on Meteorites and Krinov rose to the post of Learned Secretary. On the strength of both field and academic work he was probably the best single authority on Tunguska at the time. The asteroid 2887 Krinov was named in his honor.

CHAPTER 8. OTHER EXPLANATIONS AND PERSPECTIVES ON THE TUNGUS EVENT

With a puzzle this frustrating and fascinating, many people have had fun proposing solutions. One of the easiest "theories" to reject is that of the mini black hole, if for no other reason than the absence of an exit wound elsewhere in the world, nor does a specific entry point exist. There is neither any seismic evidence that it penetrated or was absorbed in the Earth's core, an outcome contrary to the physics of such singularities. That would be unlikely considering the gravitational fields associated with even small black holes. The statistical chance of encountering such a mini black hole is far more remote than that of a rogue asteroid or comet.

Supermini black holes as tiny as specks of dust do hypothetically exist. This proposal, generally known under the banner of the Jackson-Ryan theory, was constructed at the University of Texas in 1973. It assumes a very high speed of approach giving rise to a deep blue trail of ionized particles in transit downward, suggesting 25,000 miles an hour. The energy released on contacting the ground does not form a crater. The solely radial damage to the forest is more difficult to explain. Ground waves only were propagated across Eurasia from the fall site. Did we encounter the passage of a microscopic singularity wandering through the cosmos as a tiny remnant of the Big Bang or the final imploded core from a star following a supernova? Those are the origins of black holes consistent with cosmological theory.

As noted, the seismic disturbances show a purely surface wave action. Despite searches there is no corresponding abrupt earthquake, event or tsunami recorded elsewhere to indicate that a black hole singularity reemerged from the surface of the Earth or sea bottom. The theory proposed an escape point some 1,000 miles east of Nova Scotia. Neither does the mini black hole idea account for the huge dust legacy that followed the projectile. Most physicists agree that a collision between the Earth and such a naked singularity would have caused more cataclysmic damage than 830 square miles of flattened and singed taiga.

How about an inactive nebula? The active appearance and final energy released by the TCB hardly fits such a notion and the term "nebula theory" has not entered the list of alternate possibilities. Plekanov's "dense cloud of cosmic dust" gained very little acceptance as a theory in 1962, the hypothesis historically originating as a suggestion from the Belgian De Roy and the Russian Vernadsky.

Anyone for mirror matter or anti matter asteroids? Any such matter entering the atmosphere would have reacted at a far higher altitude. It would respond to the most tenuous of regular matter and probably to a soft vacuum. Anti matter is a huge rarity besides and is not believed to exist in the solar system.

The American astronomer Lincoln La Paz first proposed an anti-matter solution as early as 1941 and did other work on the Roswell and Socorro incidents. Such an asteroid would be rare to the point of *sui generis* and the odds of it crossing the Earth's path very low. The existence of mirror matter is a debate for astrophysics. Colliding black holes and gravitational waves in spacetime are like a playground for theoreticians. Like the mini black hole scenario, the devastation to Earth would have been far more profound than incinerated trees and herds of reindeer in the wastes of Siberia. Bursts of pure energy of this hypothetical sort are in the realm of cosmology. They are far above the levels of theoretical science we employ to seek a pragmatic solution to our mystery.

See the "Cowan-Alturi-Libby theory" 1965 for a full hypothesis of anti-matter for Tunguska. The atmospheric radiocarbon predicted to form has not been located. To the further detriment of the proposal there was no known gamma ray production by the conventional matter of the planetary medium being annihilated by the approaching anti-rock. The differences between anti-matter and mirror matter are purely hypothetical pursuits. There is no evidence to suggest that they or mini black holes are on the rampage in our cosmic neighborhood.

In the annals of Tunguska pseudo studies a crank element is present. One preposterous story suggests that Nicola Tesla was experimenting in New York and unleashed a destructive wave with his "Wireless Power Transmitter." His workshop was not even functioning at the time.

A super massive bolt of ball lightning somehow descending from a cloudless sky is another fringe notion that shows a fundamental lack of understanding of electrical phenomena. Presumably produced in active storm clouds, these luminous, small spherical electrical quirks are controversial to exist at all. There is some evidence but little explanation for ball lightning that can occur on small scales during ferocious storms.

St. Elmo's fire is a lot tamer, more common and less terrifying in appearance. It is also more readily explained as an electrical weather phenomenon in which luminous plasma is created by a coronal discharge originating from a grounded object in an atmospheric electric field. It was commonly noticed at the metallic tips of masts of ships at sea during storms and is named for the patron saint of sailors. It has been regarded with mystery and numinous awe in the past.

The natural H-bomb proposal of D'Alessio and Harms in 1989 mostly requires a deuterium rich comet as a delivery vehicle complete with nuclear fusion explosion on arrival. Deuterium is a natural heavy isotope of hydrogen with a natural abundance in Earth's oceans of 0.015%, making it an important but serious rarity here. It also requires a mechanical or kinetic explosion to trigger a thermonuclear reaction. There were enough bangs at Tunguska to go round but the lack of concentrations of radioactive isotopes left over is inconsistent with what should be found in the wake of a nuclear explosion. In 1990 Cesar Sirvent also proposed a deuterium-bearing comet as a model for the energy source. But heavy isotopes of hydrogen or any such components are not thought to comprise any part of cometary bodies. Their low mass and cold origin are quite contrary to the idea. Whether deuterium was produced solely in the Big Bang process is a scientifically purist idea but we cannot place any significant source of it in the Oort Cloud, whence comets come. The temperature and pressure conditions were also insufficient to trigger a nuclear reaction. Uranium compounds are equally non-existent in the regular comet or asteroid by their known compositions. We accept that they are not all alike but neither are they that unalike as a class of body.

How deep a personal interest in Tunguska was taken by the pernicious Stalin and the KGB is a matter of debate. There may have been a military expedition to the site in the late 1940s; if so, it remains shrouded in state secrecy to this day concerning its findings or conclusions. Some voices insist there was such covert activity to the extent of a secret military base temporarily set up in the area. On

4 October 1959, the *New York Times* relayed an article stating that the "Russians suggest blast wave in Siberia in 1908 may have been a gigantic nuclear explosion of extraterrestrial origin." Apparently the spaceship and atomic heresies briefly joined forces for the official opinion.

Unraveling Soviet cold war propaganda is a subject in itself. Whether or not they were better informed, equipped and accomplished than their Western counterparts in science, technology and political ideology, conveying that impression was a defensive technique for decades and was useful to the US as well as a means of stimulating interest and funding for our own developments.

Plenty of good academic work on Tunguska has not been translated from Russian and we welcome the newly available English-language *The Tunguska Mystery* by Vladimir Rubtsov (2009). This contains some illuminating summaries and informative lists of references. It does not appear that the Tunguska mystery was fully solved and entirely suppressed years ago by the Soviet authorities. Any such revelation would likely have come out after all this time.

The Tungus event has entered many SF plots. "The Blast" by Kazantsev as a tale interwoven with fact was the original, published in the Russian science adventure magazine *Vokrug Sveta* (translated as "Around the World") in 1946. It invoked an alien space vehicle seeking water from Lake Baikal, whose mission tragically came to grief.

For those interested in rational analysis, the principal evidence remains the roughly 700 eyewitness accounts preserved. Kulik distributed about 2500 questionnaires and sought information by invitation in regional newspapers. In writing and speaking on the phenomenon, he remained convinced of a meteor cause and the importance of locating the body that was valuable in several ways. As noted, most direct reports concerned great noise and ground tremors rather than sighting the mysterious flying object. A bright ball or cylinder with a glowing tail is the consensus when we examine the statements.

One outstanding oral recollection recorded on film years later is from the local peasant Svetlana Polonov, who was about eight at the time. She remembers being by the river with her father that day. She recounts clearly seeing a "chimney stack on its side" of a blazing object that altered direction and briefly headed back up in the sky. On the explosion she shielded her face with her hands and by the intense light clearly saw the bones of her fingers. Mysterious illnesses and deaths from unknown causes subsequently occurred among the people she knew, according to her testimony.

Roy Gallant's courageous hands-on journey into Tunguska and resulting book places the eyewitness accounts on the map in detail both within and outside the fall zone. An old man named Lurbuman is said to have died of shock, apparently on hearing about the destruction to the forest from his son, Ulkigo. The psychological trauma and destruction to property are commonly spoken of in the homespun accounts. The individual with the broken arm later succumbed. So we must revise our statement that the Tungus event killed nobody but reindeer and wildlife at the time. There are, apparently, two human fatalities associated and rumors of more deaths subsequently.

Whether the TCB changed course as it came down has been the subject of lively debate. Ultimately we can only agree on its very final trajectory. There are too many possibilities to draw a fully reliable path. It can be inferred that it followed at least an arc in direction.

Any mutations appearing among cattle and livestock or subsequent births involving undesirable genetic effects would have been summarily killed by the Evenki. They would have seen this as a judgment and punishment in light of their religious beliefs. There is no indication that this occurred, but they had no incentive to advertise it if such if it did occur. The references to boils may be attributable to smallpox rather than an unknown cause of disease. There was an outbreak there in 1915 of only too familiar etiology. The scabs on the cattle may simply have been burn injuries from trees ablaze, a straightforward explanation. Scientific papers published in the mid 1970s by the USSR Academy of Sciences referred to mutant effects in the tree population along the trajectory ostensibly taken by the TCB. As we noted from 300 soil samples and 100 plants examined in 1959 by Plevhanov's large teams, there was up to twice the level of normal radioactivity at the center as measured by them. The faster growth rate of trees and vegetation in the region had been discernible early and was clearly mentioned in Kulik's work.

However, the basic condition of there being human witnesses available for comment in the 1920s obviously shows that no great mysterious killer epidemic or radiation-based fatalities had taken place. They were still living to relate their impressions. There is also no reason to doubt the honesty of the words of Polonov. The subjective memory of an individual many years later and who was a child at the time can be psychologically challenged but there is no reason to question the integrity of her account. Deaths directly associated would be the hardest of cold facts recalled under any state of mind. There cannot be anything subjective in the alleged demise of so many villagers but this is solely according to her account. There is no reliable record of deaths following and attributed to the event and a local epidemic of more familiar disease was a more likely cause. In the very limited number of cases of human bones exhumed and evaluated from individuals approximately present in 1908 there was no abnormal radioactivity found. No medical descriptions or reports for witnesses as a patient population were ever done. Here is another approach beyond reach and closed to us.

As Rupert Furneaux concluded in *The Tungus Event* 1977, no single explanation for the mystery is wholly satisfactory and we have as yet no elegant answer. In human history the Tungus event has been described as too late to become lost in folklore but a little too early to be adequately recorded.

One reason it would be helpful to have a better understanding of what happened is this. In descriptions of the event, one repeatedly finds the grim warning that a similar event today could be grossly but easily misinterpreted as a nuclear attack and precipitate a major war of retaliation. And, in any event, the fact remains that whatever it was could conceivably happen again, and perhaps not in the frozen wastes. We only need to look up at the Moon to remind ourselves of the episodic cratering and great bombardments of the past.

Let's return to the big picture for an examination of planetary neighbors. Major craterings are quite apparent on the older and non-reworked faces of the other planets and their satellites. The planet Mercury, or Jupiter's moon Callisto, the third largest satellite in the entire solar system, are thoroughly battered and cratered.

The single most spectacular lunar surface feature, the Orientale Basin, is not visible from the Earth except for a hint of a great walled rampart under ideal conditions of lunar libration. One gets a peak of a collective 9% of the other side of the Moon over time. Our whole generalized and romantic image of the Moon would have been subtly different if it were on show. Besides the bluish mare basins, the crater Tycho is probably the best-known naked eye feature. It is a bright and comparatively recent impact feature. The progress of civilization was not conducted beneath a cold watchful eye in the sky that would, no doubt, have been one traditional interpretation of the Orientale Basin's appearance had it been on permanent view.

On improved techniques for studying meteors striking the Moon, the surface based seismometers deployed by the Apollo astronauts detected the impacts of presumed Beta Taurids in June 1975. Simultaneously, there were small disturbances in the Earth's ionosphere showing that a slightly greater ablation of small meteors than normal occurred in that brief period. We are confident that in the present epoch only the periphery of the greater and convoluted Taurid streams touch the Earth's orbit. In practical terms they form a very minor stream.

Incidentally, it has been said that nothing falls on the Earth. It falls towards the Sun and the Earth gets in the way. Bear with me and we might encounter a unique exception to that rule of celestial mechanics.

That a unique geothermal gaseous explosive anomaly from within the Earth took place is not consistent with the observed descent of a body from space. An eruption of methane in a ground-up explosion clearly is not suggested by thousands of smashed and charred trees clearly not uprooted in some upheaval from beneath. Such things have however occurred as a natural phenomenon on smaller scales.

In 2002, the scientist Wolfgang Kundt constructed a theory of natural gas proceeding upward from volcanic vents that created a methane fireball. Tunguska as a geophysical region is actually a part of a large igneous province geologically formed at the time of the Permo-Triassic boundary. Such heating and buildup of gas beneath the continental lithosphere finalized with an explosive release of some 10 million tons of natural gas causing the Tungus event. It could have been the present day formation of a kimberlite. A verneshot is a hypothetical volcanic eruption caused by a buildup of gas deep within a craton which form the old and stable part of the lithosphere. It is named after Jules Verne and suggests that mantle plumes cause heating and the buildup of carbon dioxide beneath the crust.

There is a clutch of entirely terrestrial explanations. A major electrical interaction from the crust up to an ionospheric anomaly in the much lower reaches has been put forward. Some even do away entirely with a TCB itself in the face of all accounts of the approaching object. Geometeors have no established scientific basis. Hypothetically, they are interactions between atmospheric and tectonic processes but the entire notion is placed in the crank category.

Unraveling what the local people saw that morning is fraught with difficulties but they unequivocally observed and more commonly heard something from the sky. There are no simultaneous earthquakes, volcanic eruptions or local hot springs to consider. The contemporary data was vague but not that inexact. To the detriment for these geological and terrestrial attempts, there is no account of a gaseous aftermath or upheaval to the local crust from below. The former would have been clearly detected in soil and plant samples and the simplest terrestrial surveys.

The tectonic theory or some relation of major changes in atmospheric pressure triggering an earthquake is pseudoscientific. It was proposed as an explanation by such enthusiastic independent researchers as Dr. Andrey Olkhovatov on his website. The shockwave above the taiga was equivalent to 5.0 on the Richter Scale (introduced in 1935 to quantify the amount of seismic energy released in an earthquake) and was detected on seismic stations across Eurasia. Eastern central Siberia is not an earthquake zone of that magnitude as has been long established. Our event was not seismic in cause but was so, somewhat, in effect. There would have been shakings of the ground elsewhere in Eurasia of detectable and related scale. These alternative, geothermal, volcanic and marsh gas speculations are fun but I agree with Robstov that the only contribution these models make is a negative one. We are contending with a cosmic unknown and commence with unusual meteors and comets. Geophysically, the Siberian craton is thought to result from the Pangea 1 Paleoproterozoic super continent with mountains of Caledonian age. We will not make progress by uninventing our profoundly mysterious space body. Dr Olkhovatov devoted a book to a purely tectonic explanation.

Returning to stricter scientific methodology, the best test of any hypothesis is to successfully predict empirical data. Karl Popper reminds us that every genuine test of a theory is an attempt to refute it. We must, as a scientific community, press on with the investigation. A century is too long to still be wondering from our supposed enlightened standpoint and find no definitive answers. The matter of Tunguska has been declared solved and closed on several occasions.

Best recommended for ongoing Tunguska research is the University of Bologna Web Pages. They might be our premier resource for uncovering new data or successfully reinterpreting the existing facts. For example, their deduction from the revised Fast data on the orientation of fallen trees integrated with the 1938 and 1999 aero photo surveys suggests that the TCB may indeed have been a multiple bolide. Two or possibly three bodies closely spaced and on parallel trajectories were possible. This could be an elaborate solution combining the observations of the trajectory or trajectories and making them fit at last. Victor Konenkin, a contemporary schoolteacher from Vanavara, discovered that the TCB was seen both to the east and south of the Great Hollow, a fact we cannot otherwise interpret. The first had the greatest mass and released the most destructive energy on the forest below.

The Italians last visited the fall site in July 2009. Tree and tree core samples were taken around Lake Cheko, methodically pursuing further magnetometer readings along the lake bottom. There are no known meteorite fragments on the lake floor. That latest press release described that a small anomaly on the bottom of the lake may have been detected but the following day was not in evidence. Sometimes we must resign ourselves to the condition that the trail has long since grown cold. Alternatively, techniques are continually improving, sometimes momentously. What might be technically possible in the future sifting of evidence may deliver the information we seek.

Other nations may have data to contribute. At what location and whose geographical zenith did it first appear? There must have been some visible display before that first sighting north of Lake Baikal. Projecting back along its assumed path, what was there to be seen as a high fiery trail traversing eastern Mongolia, Korea, Japan or the western Pacific Ocean and how many minutes previously? Was it speeding up by the acceleration due to gravity as it plummeted or slowing down as it shed mass and encountered atmospheric friction? Was it brightening or dimming intrinsically? These are easy descriptive factors that escaped being written down. It seems logical that the TCB played to a much wider audience than ever filled in the questionnaires or whose words are on the record.

That suggestion of a body in passing might expound the problems of no crater and zero physical vestiges but in all the first hand narratives there is no indication of any part of the body bowing back out into space.

The Grand Tetons meteor of 1972 did that very thing in slicing through the Earth's atmosphere then going on its cosmic way, mercifully without airburst or impact. The unusual sight in the daytime sky of a big passing meteor complete with tail was partly captured on film from ground level. Estimate show that the Grand Tetons body was travelling at some 900 miles an hour at a height of 36 miles in those moments. There is a clear case in point of a meteor remarkably passing through. There would be too much air braking friction at the height of a low 5 miles to allow such a passage to take place, the altitude where the Tunguska explosion occurred.

Curiously, in 1985 Samuel Sunter of Victoria, Canada looked back at an experience when he was nine years old, in Northumberland, England. Here is the quote from the chapter "Out Of The Blue" which investigates several deeply mysterious shocks from nowhere researched for the "Mysterious World" series.

I saw, looking northeast on June 30th 1908, a large red ball of fire about three times the size of a full moon. It looked just like a hole in the sky. On the other side of the hole it looked like flames, just like looking into the fire box of a locomotive. But what made me afraid was a solid beam of light which reached right down to where I was standing. This made me afraid and I ran into the house. So I do not know how long it lasted after I first saw it. Even today I have a very vivid memory of it.

It would be gratifying to discover some missing links of contemporary eye witness accounts, geographically removed from eastern central Siberia but at the exact corroborated time. Some contribution like a sighting from an area SE of northern Lake Baikal slightly earlier in the dawn could be invaluable.

We must show grave skepticism, however, that this narrative was first expressed eight decades after the event. It is timely but lacks any equivalent testimonies at all. Subjective and distorted memories over a lifetime are psychological home truths. The account exists in complete isolation to the placement of any other observers and from a vantage point 4,000 miles west in northern England. This is enormously more than the curve of the horizon permits, of course. Apart from a location across "the roof of the world" for our observer, we entirely assume that the TCB emerged from the SE. Decisively, there are no sightings from the densely populated zones of Russia or Europe to the west of the fall site. Did he see some impression of the prime blast or part of the immediate aftereffects of the explosion? Did the nine-year-old reliably witness anything real and relevant? Without photography (this is absent from any Tunguska studies until Kulik) or corroboration this is notoriously unreliable and our scientific court rules it inadmissible as evidence.

How much of the bright night displays and state of the atmosphere in early July 1908 was caused by the incineration of the object itself? We assume a major part.

What proportions in relation to a large cometary tail, forest fire ashes, raised soil and rock particles and burning debris caused by the ground explosion is impossible to clarify now. Probably it was a low factor apart from the immediate environs because the forest was damaged but not wiped out or fully consumed in the process. We assume that the dissipated residue was a product of the TCB itself. The material and particulates placed in the atmosphere was, of course, substantial. We could have managed a late night round of golf those evenings. Some farmers literally made hay while the Sun shone with the abnormally bright twilights so widely seen.

There are many ordinary looking photographs of regular terrestrial objects taken in early July 1908 where it is oddly light at the late evening. Following the grand appearance of Halley's Comet in 1910, Max Wolf of the Heidelberg Observatory later suggested that a great and dusty cometary tail had descended over Siberia and western Eurasia two years prior. He was not informed of the Tungus event at the time he made the proposal and the strange lack of darkness was affecting observations there. Wolf was a contemporary authority on asteroids and comets with a robust record as an observer.

As a dedicated amateur astronomer I was determined to come up with some viable natural explanation and have done so, albeit unlikely and without precedent or support in solar science. Further, I acknowledge that I subsequently found it not wholly original. As much as I would have liked it, I did not formally incubate the solution we will now explore together.

When it comes to science by democracy, S. Verma's book quotes a Russian website's results of a poll on the cause of the event. Educated opinion here places 31% in favor of a cometary explanation, meteor or asteroid 27%, alien spaceship 9% and the remaining 33% by causes unknown. Quite simply, a comet would have fizzled out much higher up, a big meteor would have left pieces on the ground and there is no direct evidence to suggest that ET dropped in with a bang. Even informed and considered opinion has a third of its sum total researchers still gripped by mystery.

Norton's work seems convinced that a stony-iron meteor, conclusively 7 million tons in mass and 500 feet in diameter entered at 7 miles a second and on reaching the middle troposphere exploded in an instant with colossal energy. Incidentally, the troposphere is the lowest atmospheric layer that at middle latitudes is about 11 miles tall, shallower towards the poles. It contains 75% of the mass of the atmosphere and 99% of its water vapor and aerosols.

Whatever type of cosmic intruder we opt for, there is the matter of the "dust" residue for the days following that no model has successfully accommodated. The chemical constitution of that wave of material remains unknown. Scientific models only need apply and we shall pass over time warps and paranormal causes.

1. Meteor or larger asteroid.

The earliest and traditional assumed case but lacks physical evidence like a crater or meteoric strewn field. A body sufficiently dense or protected from the stresses of descent to the height of the explosion should leave some direct remnant. The Beta Taurid meteors climax annually on June 30. An increase in the iridium concentrations at the icecaps might stratigraphically relate to 1908. This could plausibly have been delivered by asteroid fragmentation. Whether a huge meteor could pack such a punch of energy is a matter of controversy.

2. Comet or cometary fragment.

From the known rocky, icy composition of comets and based on laws of dynamics it is remote that such a conglomerate could survive down to an approximate 26,000 feet altitude. Some explosive component is deemed unlikely although not all comets are alike. A slow descent or protected mini nucleus/fragment type of proposal could allow the close approach to the ground before erupting but nothing was seen coming from outside the atmosphere. No tail typical of a comet was seen at any stage.

3. Mini black hole.

The lack of exit or entry wounds, the strictly surface waves for the first "meteor" ever seismically detected, but above all, the absence of any exit phenomena of a meteorological or geophysical nature anywhere else on Earth negates the theory. Searches to locate such possible records have been made. The effect of a mini black hole would have been far more radical than the Tunguska fireball and associated forest damage.

4. Anti-matter.

A highly theoretical entity in itself, the hypothetical reversed charges and polarities would have neutralized in contact with the regular matter of the upper atmosphere. A limited gamma ray burst would be more likely in such a case. There is no signature of any pure energy burst.

5. Mirror matter.

Possibly distinct from anti-matter, and a theoretician's paradise, it is difficult to advocate how this could have entered the solar system let alone the domains of Earth without being snuffed out at the quantum level far earlier in its career.

6. Underground volcanic venting/geometeors.

The evidence to support an upheaval from beneath would be clear geologically, and it simply is not there. This also ignores the celestial fireworks of the observed TCB from hundreds of witnesses over thousands of square miles. There is no geological case for a massive electrical interaction between the ground and a strange celestial object.

7. Lightning ball.

The unexplained anomaly of curious and ball lightning does exist. We must not ignore phenomena we cannot explain. But a maverick and giant bolt of lightning is not a viable suggestion for a clear, dry summer's day with no stormy conditions or related meteorological activity associated. A huge electrical interaction between the TSB and the ground is similarly untenable.

8. Spaceship hypothesis

The most romantic of ideas, sadly lacking any direct evidence, is that a crew of intelligent beings from the cosmos crash landed a vehicle after a failed attempt to touch down on Earth. It retains popularity mostly as an entertainment feature that has risen to a cultlike status over time.

9. Plasmoid body from the Sun.

The notion of a plasma energy entity arriving from the Sun or building up in the near magnetosphere of Earth and dropping that day was specially designed to explain "traceless Tunguska." In this context a "plasmoid" is conjectured as a coherent structure of plasma and magnetic fields that could, very speculatively, account for ball lightning and, less uncertainly, magnetic bubbles in the Earth's magnetosphere. Also, features in cometary tails, the solar wind and the solar atmosphere may have this root cause.

The term was first coined W.H. Bostick (1916–1991) to describe a plasma magnetic entity. Plasma is a partially ionized gas in which a certain portion of ions is free rather than bound to an atom or molecule. It is considered a separate state of matter along with solids, liquids and gases. It was first identified in a Crookes Tube in 1879, referred to as "cathode ray matter" by J.J. Thompson (1856–1940) and termed "plasma" by Irving Langmuir (1881–1957) in 1928 due to its resemblance to blood.

It is very common in the universe but hardly to everyday life here on our pale blue dot in space. Stars are made of it. The spaces between them are thinly veiled in an interstellar medium (distinct from the hypothetically debunked all-pervading ether.) The ionosphere is a shell of electrons and electrically charged atoms surrounding the Earth from a height of 30 miles to over 600 miles altitude. Ionization depends on the Sun and its activity and goes on continually over numerous layers high above the atmosphere.

Could some rare solar sub-coronal ejection of magnetically stable shielded plasma reach us from deep within the Sun to enter the atmosphere and dynamically neutralize itself in a 10–15 megaton explosion 5 miles up? The Russian researchers Zhuravlev and Dmitriev first sketched a plasmoid hypothesis in 1984.

The likelihood of any such aberration forming, escaping the Sun's gravity and traversing the 93 million miles to us on target is at best remote. An interstellar plasma package is even less likely to survive the colossal distances and arrive at Earth still active.

The solar wind is a vast and varying stream of charged particles put out into space continually from the Sun. The known behavior of the Earth's magneto-sphere in response to solar flares as aurorae and magnetic storms have never given rise to bolides, explosions or impacts directly. Acoustic waves obviously reaching to the lower troposphere and a crackling sound associated with an au-rora have been observed. There is no known emanation from the Sun to account directly for the Tungus mystery. Space weather is not that stormy.

How about some powerful escalation of a discrete self contained energy pack-age in the Earth's magnetosphere that sensationally neutralized itself in an arc to Earth and eventuated in a fierce, self-engulfing immolation? Could we ever relate the path of the TCB to a magnetic field line on Earth? Was it heading Magnetic North?

A hypothesis that a protected fragment of the Sun's internal hot plasma was somehow hurled toward Earth to both interact with the atmosphere and almost reach ground level with magnetic disturbances was also suggested by Alyona Boyarkina of Tomsk State University in the 1990s. Zhuravlev is quoted saying that this was a cosmic object the composition and structure of which is un-known to astronomers and physicists.

On 27 June 1908 and for several nights following, Germany's Kiel University had recorded small periodic deviations in the observatory's compass needle. Was this indicative of a highly magnetized yet presumably small, dark and unseen body approaching the Earth? A similar effect was noticed after the event in the compasses and magnetometers at Irkutsk, where the geomagnetic storm was re-corded for about five hours. At Pavlovsk, near St Petersburg, and west in London there were also minor but unexplained interference in the direction of compass needles.

The *Times* of London for 3 July 1908 gives a curious column:

> Mr G.J Newbegin drew attention to the disturbed state of the solar
> atmosphere, showing a drawing and giving a description of a very large
> prominence that he had observed and measured in the morning of that day
> (July 1) and that showed unusual changes of form. Allusion was made by

Mr E.W. Maunder and Mr H.P. Hollis (both of the Royal Observatory) to the long lasting aurora of the previous evening.

The following day and in the same publication, we read concerning "The Recent Nocturnal Glows."

The remarkably ruddy glows, which have been seen on many nights lately, have attracted much attention and have been seen over an area as far as Berlin. There is considerable difference of opinion as to their nature. Some hold that they are auroral; their colour is quite consistent with this view and there is also the fact that Professor Fowler of South Kensington predicted auroral displays at this time from his observations, which showed great disturbances in the Sun's prominences. There was a slight but plainly marked disturbance of the magnets on Tuesday night and this materially strengthened the auroral theory as the two phenomena are very closely correlated. However, this was shaken on the following night when the glow was quite as strong but the magnets were exceptionally quiet.

The E.W. Maunder (1851–1928) referenced in paragraph one was the noted solar observer who discovered the little ice age that took place 1645–1715 within the major solar cycle that became known as the Maunder Minimum. In the record of sunspot groups and their appearances is revealed an 11-year cycle for the Sun. This is not news. Secondly, what sort of equipment were Newbegin or Fowler using to make any such observation reliably and singlehandedly in 1908? Lastly, nowhere else do we note any major solar prominence, flare, sunspot grouping or observed upheaval at the time to corroborate these observations. It could be that a very major prominence is connected and was seen by these observers alone. Sadly, no photographs are available. The meeting minutes for 1 July 1908 of the British astronomical Association includes mention of a "very fine and bright prominence at the limb" (of the Sun) and that Newbegin had achieved some good timed solar work over recent days. The night time luminosities and sunset effects shifted rather to the north are also mentioned. By calendar this discussion of 1 July was two weeks after the phenomenon of 17 June 1908 in Tunguska, note not the day after the event. The minutes include that "curiously enough, about 15 days ago in the same (solar) latitude, they had another very fine prominence on the eastern limb." (*Journal of The British Astronomical Association* Vol XVIII No. 9. These records of the minutes of the meetings in 1908 available on line.)

Perhaps in the light of contemporary solar physics with such assets as SOHO in orbit and with so many coronal mass ejections observed this could be ascribed to a "halo event." No instrumentation capable of this existed in 1908, of course. A coronograph which places an occulting disc in the telescope's optics to effect a artificial "eclipse" of the Sun's image would not be equal to the task.

The *New York Times* for 3 July 1908 may have accidentally hit the nail on the head, noting that the remarkable lights in the northern heavens were due to important changes in the Sun's surface, causing electrical discharges.

The TCB is generally agreed to have approached rapidly at a low 30º to 35º angle to the ground or possibly less and more likely from the SE, possibly with a change or two in direction before detonation. Plausibly it had an azimuth of 115 degrees along its final trajectory and even veered upwards shortly before the explosion. It can also be asserted that it came directly from the position of the known radiant of the Beta Taurids. By another observer's subjective testimony it broke away from the Sun. This must not be over interpreted, as ever. Information concerning anything occurring on the Sun would take at least eight minutes to reach Earth across 93 million miles at the speed of light. In that time the Earth has covered about 9,000 further miles in its orbit.

The solar wind takes several days to get here as a continuous torrent of particles. The time lapse between a solar flare visible on the Sun's surface and subsequently reaching us to interact with the Earth's magnetic field is understood. It helps predict the interference with radio communications with a few days advance notice after a major flare erupts on the photosphere. The phrase "space weather" has been minted to describe the changing environmental conditions in space. The website spaceweather.com even gives a daily update on the speed and density of the solar wind.

The 11-year sunspot cycle has been reliably monitored since 1755 by simpler means of observation. We were midway between the recorded February 1902 to August 1913 solar cycle at the time of the Tungus Event. Mercury, Venus and Earth were in a conjunction in those weeks with practically minimal distances between them, not an altogether rare alignment.

In conclusion, we ask, could it have been a truly unique discrete magnetically-shielded plasma body from the very depths of the Sun that remarkably survived to reach Earth and spectacularly plunged into the atmosphere? Whatever it was, the final fireball disturbed the magnetic field on approach and caused a localized magnetic storm on exploding but left no tangible traces of itself. That admittedly would be an implausible piece of random navigation, a shot in a million for the most rare of hypothetical solar phenomena.

More logically, a rare development in the magnetosphere over an unguessable time period could have been triggered. Perhaps there were two or more discrete projectiles released in the process, travelling closely together. As an original suggestion, could there have been some unusual buildup in the magnetosphere of Earth to produce them? Or, could the Sun have erupted a rare scattershot of energetic plasmoids, only one of which we noticed?

Alyona Boyarkina, a senior scientific researcher of Tomsk State University, was probably first to suggest that a fragment of the Sun's hot gases was thrown toward Earth. What sort of a unique body was produced is open speculation but perhaps the magnetic anomalies of Tunguska hold some clues.

Suslov's fieldwork with the Evenki elicited the fact that, in their view, a piece came off the Sun, fell on the Earth and burnt the taiga. It is strange to have come so far with high science only to find a viable description in the words of simple folk who were present at the time. In simplistic terms this could refer to the final solution and the case this book supports.

What if the "body" was neither cometary nor meteoric in nature and arrived by chance on a date coincidental to the peak of the routine Beta Taurid meteor stream? The date 30 June is either the answer staring us in the face or a banal coincidence of timing. The sunward direction of origin is therefore significant. For some reason a hefty dust residue was created by the major heating and ionization along its final atmospheric path but it possessed only a very small material nucleus or inner structure. Scholars are generally agreed that this was an extraordinary object and relatively compact in size. The speed was less than cosmic velocity on arrival because it was seen and heard for some minutes. The most plausible scientific explanation then is that the nature of the TCB was an indirect solar plasmoid.

It came out of nowhere. It hotly blew itself away with a giant residue of dust aimed upward and laterally westward. It had an inner source of energy far more powerful than forces acting externally on an asteroid or cometary core. There is no association with Comet Encke or any other recognized longhaired star. Cometary tails can become detached, as we have decisively observed, but they are far too low in mass to explain the explosions and temperatures associated. Earth has walked right into cometary tails in the past and nobody even noticed.

Paradoxically, some bright, blazing, non-smoky tail seems to have accompanied the Tungus projectile that was more commonly described as cylindrical or spheroid in shape and very bright. There was no overbearing great mass of dark dust trailing behind. As Robstov noted, the Sikhote-Alin event followed the rules of known meteor science with obliging craterlets and physical debris for investigators to enthusiastically gather up. The Tunguska object absolutely did not play by the rules.

Let us return to Voznesensky, Director of the geophysical observatory at Irkutsk, he denied himself the mantle of becoming the founding father of Tunguska studies. He was far more informed in 1908 than he revealed at the time, limiting himself to seismic reports and earthquakes. His findings were not published until after the wars and the easing of immediate tensions in the first years of the Soviet Union in 1925. Academics might wish he had shared his findings earlier, but even in 1925 he makes no mention of the geomagnetic storm solely recorded by his own facility or, for that matter, the associated atmospheric effects. Rubstov suggests that Voznesensky deliberately avoided sensationalism as it simply would have been too much for the scientific community to expound at the time. His reticence to publish might have been similar to that of Copernicus and Darwin. Scientifically, Voznesensky can be forgiven the slight error of timing for the event as he had worked with the average speed of seismic waves. He overestimated the height of the "rupture of the meteorite," too.

Among more modern studies of geomagnetic effects rendered by thermonuclear explosions researched by Plekhanov and Vasiliev, a brief entry in a contemporary German scientific journal contributed by Prof Weber of Kiel, located in 1959, was of the greatest interest. It succinctly reported a strange disturbance of magnets over a few days shortly prior to the event. The effect repeated each night from 27 June through 30 June. The effect ceased following the event. In Weber's own words from the Physical Institute of Kiel University on 11 July 1908 to the editors of *Astronomische Nachrichten* (Astronomical News):

> In the course of the last 14 days, the photographically recorded curves of magnetic declination showed no disturbances of the sort that usually accompanies the Northern Lights. But is should be noted that several times and indeed all the time over several hours, there were small uninterrupted vibrations of magnetic declination curves of about 2' amplitude (angular minutes) and 3 min period which could not be traced to known causes (e.g., a streetcar vibration) these as yet unexplained disturbances took place:

June 27–28, 6:00 PM to 1:30 AM
June 28–29, the same

June 29–30, 8:30 PM to 1:30 AM

Author's note: Careful but straightforward analysis reveals two identical effects of 7 hours 30 minutes each over a dual twenty-four hour period, then a briefer 5 hour 30 minute anomaly that commenced later but concluded at the same diurnal time. The originals of these magnetograms when sought out were determined to have been lost in the War.

In February 1960 the geophysicist Kim Ivanov of the Irkutsk Geophysical Observatory (renamed from the Irkutsk Magnetographic and Meteorological Observatory) had mailed a package referring to the magnetograms preserved in the old institution's archives from 1908. He was confident that some record of the event was preserved in the disturbance to the geomagnetic field following the explosion that day. It was preserved in the archives of the old institution. Ivanov did not enter into theorizing. The studies discouraged speculation that the magnetic effects were related to disturbances caused by typical meteor bodies in flight. The magnetic effects of meteors are mere seconds in length and strangely, there was no sign of the normal meteor effect when the TCB was in flight, curious in itself for such a body presumably much larger than a typical meteor. The resemblance to a regional geomagnetic storm recorded there was distinct. Curiously, they resembled the artificial geomagnetic effects resulting from the nuclear testing in the Pacific Ocean atolls two years previously. A detailed study was published in 1960. Two senior ITEG members made enquiries to several observatories but there were no other magnetograms showing anomalies available.

Closely interpreting a 1996 paper from V. Zhuravlev on the matter, the short conclusions are that a local geomagnetic storm was recorded at Irkutsk some 600 miles from ground zero and that was the sole locale. Conditions that might correlate with a chemical explosion look increasingly unlikely as the duration of the magnetic storm caused by the Tunguska explosion could not be of more than 10 minutes duration and yet the actuality was a situation of 4 or 5 hours in length. The cometary hypothesis is also weakened with the remote exception of one containing an unknown source of plasma of high density. Comets, as we have seen, are not all identical but they are low mass and not thought to have this composition. A cometary-caused geomagnetic storm would have been far more widespread whilst the strictly local nature of the Tunguska case is clear. A high temperature ionized gas suggests itself repeatedly.

No combination of a blast wave or ballistic wave satisfactorily explains the geomagnetic anomalies whilst the hard radiation of the fireball is clearly linked to the phenomenon. There was also an estimated two-minute time lag between the instant of the explosion and the commencement of the geomagnetic storm as calculated by Ivanov. The figure has since been reevaluated to be as long at 6 minutes 23 seconds. Zhuravlev proposed that the fiery explosion required this time to ascend and allow the hard radiation to interact with the ionosphere and produce the strong but local geomagnetic effect. Further ionization of the ion-osphere directly above the explosion may therefore have been involved. There could well have been a temporary hole in the ozone layer. Then the radiation stimulated electrical currents there, giving rise to the geomagnetic storms. They can also be seen as both too long and too strong to have a nuclear cause.

The normal geomagnetic storm is due to a surge in the speed of the solar wind. The Irkutsk magnetogram differs in several ways to a normal solar geomagnetic storm. Of the four stages inherent, the first entry, the phase of rise, the phase of fall and the phase of relaxation, the geomagnetic effects were overall much quicker. If a regular solar storm had been in progress that day, other observatories would have detected its presence and aurorae seen in the skies. Equipment and instrumentation were sufficiently sophisticated in the era. Instead, any other observatories must have been too far afield to do so. Even a brief solar storm from the state of the art equipment would have noticed it, and the source observatory at Irkutsk was well equipped by the standards of the day.

Finally, residual magnetization exists in the soils in the area of the Great Hollow forming a paleomagnetic anomaly. Its original source can only have been an abrupt and powerful magnetic field. The TCB was clearly in possession of an intrinsic source of energy including the magnetic. Measuring the actual field strength at the time, like so many other factors is beyond our technical grasp.

Tunguska remains an obscure place to this day. Kulik was, of course, a great pioneer with his dedication and monumental work, but it would have helped if Voznesensky had communicated everything he had probed and established at a prior date. What held him back? Voznesensky had modern equipment and a wide circle of scientific communications and observers. He had made two balloon ascents over Irkutsk the previous year and was not in any sense academically conservative or personally timid. He could easily have collected more naked-eye accounts of the body in flight.

The magnetosphere is practically defined as the zone within which the Earth's magnetic field dominates over the weak interplanetary field that extends outward from the Sun. The movement of material in the Earth's liquid metallic core generates the field. As the fluid circulates it sets up an electric current. The magnetosphere was first discovered by the American space probe Explorer 1 in 1958 and Thomas Gold expressed the following: "The region above the iono-sphere in which the magnetic field of the Earth has a dominant control over the motion of gas and fast charged particles is known to extend out to a distance of the order of 10 Earth radii, it may appropriately be called the magnetosphere."

For a little more clarification, it also distends approximately 200 Earth radii in the anti-sunward direction with a tail stretching well beyond the Moon's orbit. The lunar path lies at a distance of 60 Earth radii. The solar wind is a permanent flow of hot plasma out of the Sun in all directions with shifts in speed and intensity. One of the first experiments set up on the surface of the Moon by the Apollo 11 astronauts was a solar wind spectrometer.

Structurally, the magnetosphere is dynamic with major responses to shifts in the pressure of the solar wind and the orientation of the planetary magnetic field. One might speculate that a descent from these sorts of lesser altitudes rather than a direct bolt from the Sun may have formed the TCB. This could explain the complete dearth of sightings of an approaching object rather than postulating an approach out of the Sun in direction. Energy can be stored in the magnetotail to be later released as substorms.

Certainly, the facts of a geomagnetic storm raging at Tunguska ground zero have been very late in the historical order of consideration and were not available as data to the first researchers. It only surfaced in evidence from records of the renamed Irkutsk Geophysical Observatory in 1960. The reason their chief officer has been criticized retrospectively for a lack of action in 1908 is that there are no other reports of geomagnetic disturbances from other observatories. Perhaps the effect was strictly local to the Tunguska region, unlike atmospheric pressure waves that hypothetically were available to any equipped station in the world. The same applies to seismic data in being far more widespread. It is also unfortunate that Kulik's proposal to pursue a magnetic survey was not possible at an earlier stage of the investigation, approximately 1940, and shelved indefinitely by the looming specter of war. It is also debatable that equipment equal to the task of detection would have been practically available.

Comparing the event to a geomagnetic storm similar to those produced in high altitude nuclear explosions, one can only say that the disturbance was too brief (five hours) and neither can it be explained by a typical solar geomagnetic storm. No such storm was noted anywhere else. The nuclear theory, attributing the magnetic disturbances to the release of hard radiation, is contradicted by the fact that there was a time lag over 6 minutes between the explosion and the commencement of the geomagnetic storm. Certainly the TCB was the source of a powerful magnetic field. Paleomagnetic anomalies and remnant magnetization does exist at the epicenter, looking the most chaotic around the Ostraya Mountain. Here, nearly 2½ miles from the epicenter, the magnetic field was at its strongest. The actual and quantified strength of the source 5 miles aloft, by calculation, was briefly greater than the Earth's natural magnetic field by a factor of 500 (Sidoras and Boyarkina.)

So what is the relation to the "Weber effect" — strange perturbations in the geomagnetic field that oscillated on June 27, 28 and 29 in 1908 with 24 hour separations as recorded in Kiel, Germany? If there was a magnetic precursor to the event as distinct from the massive aftereffects, that remains a deeper puzzle.

The absence of the normal, far lesser and briefer magnetic effects associated with a familiar meteor in flight further suggests a low speed throughout its path of approach. The average meteor is moving much faster than estimates related to the TCB. Dr. Victor Zhuravlev and Aleksey Dmitriev of ITEG deduced a huge and stable natural plasmoid shaped as a spindle-like magnetic bottle surrounded by an external magnetosphere. However, the plasmoid hypothesis has been designed solely to tackle the Tungus mystery. I do not think it has been proposed anywhere else. We note that nothing has been convincingly found in observational astronomy to date to support the existence of such anomalies.

The outstanding puzzlers include the great body of "dust" deposited in the atmosphere and the strange nocturnal illuminations and bright nights it caused. They seemed to increase from east to west with no significant sky glows reported in the Tunguska region itself. The effect and view to the north was unobserved and unreported. A good range of photos from those few days of ordinary objects after 30 June exist in testimony of the white nights. By examining them photometrically the intensity of the strange effects can be estimated by modern techniques. The long twilights also dissipated and faded far more swiftly than the dust input from the Krakatoa explosion with which it is always compared. But analyzing the data eighty years later, the mass present above Mt. Wilson Observatory has proven not to have been dust at all but an aerosol suspension in the air of ultramicroscopic particles. We have always rather blandly assumed it was fine particulates as debris from the body itself and of a dusty nature. Instead, we must confront some strange fluorescent substance of unidentified chemical content. The 1908 data is limited in reliable conclusion but spectral analysis in-dicates that it was not dust. How so much fine material was propagated strongly upward and westward and inexplicably not downward on the impact site or in that geographic zone remains unknown. Did the explosion affect and produce some portion of the particulates out of the ambient air layers? Is this a more ra-tional explanation than the TCB blowing itself to ultra tiny pieces? How do we square traceless Tunguska with the peculiarity of the white nights? Could there still be physical remnants lying very deep at the Great Hollow?

The radiation liberated set off a chain of ionospheric currents in an intense but relatively localized magnetic disturbance that a single geophysical observa-tory at a distance of 600 miles by chance recorded. It was the only station suitably equipped in a very limited area of geomagnetic disturbance. The 6½ second lag in its inception from the explosion remains a technical mystery within the great enigma. Hard radiation propagates at the velocity of light. Such a beam heading upward would impact the ionosphere in less than a second but there was a de-layed, if intense, magnetic response coursing downward in a very narrow pulse. A better understanding of the magnetic effects could shed some light on the greater problems.

We return to the nature of the body and the cause of the explosion. A great mass of fine particulates suspended with the perennially spinning Earth rotat-ing east to west beneath has a satisfying ring of truth to it. Calling the source icy conglomerates, volatiles and frozen hydrocarbons heating and exploding in a brief powerful flash from a piece of a comet does not fit the recipe. Neither does a big space rock.

There is no basis to believe that at the heart of the Taurid Complex lie un-known stored energy packages that go into high activity when escaping from vacuum. Meteor streams are among the most dormant and in context very minor members of the Sun's entourage. There are many regular and predictable meteor streams and those scheduled to interact with the Earth in the late June/early July period bear no distinctions. Would entering an atmosphere be a good way to activate the forces locked up in a planetary plasmoid? It seems unlikely that such entities can endure in orbit around the Sun and rest passively in a meteoroid stream indefinitely. There is not a hint of observational evidence. Let us rule out that the TCB had been calmly lodged in the heart of the Beta Taurid meteors for any period.

Whatever the truth of the Tunguska phenomenon, it retains a special place in the progress of astronomy and studies of Earth history, as we accept with humility that the universe still has mysteries we cannot confidently explain.

No matter how we look at it, Tunguska was an extremely rare event that still defies explanation despite our scientific advances. Perhaps Lake Cheko will finally give up some submerged secret and reveal new avenues for investigation. So far, this remains the greatest Earth-based cosmic mystery of the last century.

BIBLIOGRAPHY

Arny, Thomas A. *Explorations: An Introduction To Astronomy.* Chap. 10.4, "Giant Impacts". New York, McGraw Hill, 2004.

Asimov, Isaac. *A Choice Of Catastrophes.* New York, Simon & Schuster, 1979.

Barnes-Svarney, Patricia. *Asteroid: Earth Destroyer Or New Frontier?* New York, Plenum Press, 1996.

Baxter, John and Thomas Adkins. *The Fire Came By.* London, Futura, 1976.

Chaikin, Andrew. "Target: Tunguska." *Sky and Telescope.* Vol 67, No 1. Jan 1984: pp 18-21.

Clarke, Arthur C. *The Other Side of the Sky. Out of the Sun.* New York, New American Library, 1959.

Darling, David. *The Universal Book of Astronomy. Asteroids.* New Jersey, John Wiley, 2004.

Davies, Paul. *The Fifth Miracle,* New York, Simon and Schuster, 1999.

Davies, Paul. *The Goldilocks Enigma.* Penguin Group UK, 2006.

Friedman, Herbert. *Sun And Earth.* New York Scientific American Library, 1986.

Florenskiy, K.P. Preliminary Results From The 1961 Combined Tunguska Meteorite Expedition. Meteoritica Vol. XXIII (1963), Taurus Press, 1965.

Furneaux, Rupert. *The Tungus Event.* St Albans, Panther, 1977.

Gallant, Roy A. *The Day The Earth Split Apart.* New York, Simon And Schuster, 1995.

Lang, Kenneth R. *Sun, Earth And Sky.* New York, Springer, 1997.

Norton, O Richard. *Rocks From Space.* Montana, Mountain Press, 1994.

Rothery, David. *Satellites of the Outer Planets. Worlds In their Own Right.* New York, Oxford University Press, 1995.

Rubstov, Vladimir. *The Tunguska Mystery.* New York, Springer, 2009.

Schilling, Govert. *The Hunt for Planet X. New Worlds and the Fate of Pluto.* New York, Copernicus, 2008.

Semeniuk, Ivan. "Ice Age Impacts". *Sky And Telescope.* Vol.118, No 3. Sept 2009: pp 20-25.

Sobolev, V.S., ed. *The Problematic Meteorite*. USSR Academy of Sciences, 1975.

Steel, Duncan. *Rogue Asteroids and Doomsday Comets*. New York, John Wiley, 1995.

Stoneley, Jack, *Cauldron of Hell:Tunguska*. New York, Simon & Schuster, 1977.

Vasilyev, N.V. "The Tunguska Meteorite Problem Today." *Planetary and Space Science*. Vol 46, Feb 1998: pp 129-150.

Verma, Surendra. *The Mystery of the Tunguska Fireball*. Cambridge, Icon, 2005.

Voyages Through The Universe. 2000, 2nd Edition. "Asteroids and Comets." pp.. 259-277. Harcourt College Publishers.

Welfare, Simon and John Fairley. *Arthur C. Clarke's Mysterious World*. Chap 9. "The Great Siberian Explosion." New York, Trident, 1980.

WEBSITES

Hungarian Academy of Sciences. Planetologia.elte.hu/1cikkeke.phtuI?cim=Tunguska.html.

NASA Orbit Diagrams. http://neo.jpl.nasa.gov/orbits/

The Planetary Society. "The Tunguska Riddle: How Powerful Was The Greatest Asteroid Impact in Recorded History?" Amir Alexander. 15 April, 2008. www.planetarysociety.org

Sandia National Laboratories. News release 17 December 2007. www.sandia.gov

Space Weather Data. http://spaceweather.com

TunguskaHomePage.UniversityofBologna(Italy). Dept of Physics. www.-th.bo.infin.it/Tunguska

www.universetoday.com

Tungus event witness reports. www.verdalak.com/tunguska/witnesses

Wikipedia.Taurid Complex. http://en.wikipedia.org/wiki/Taurids

Wikipedia.Tunguska Event. http://en.wikipedia.org/wiki/Tunguska-even

YOU TUBE

A visit to the site of the Tunguska explosion.

The Mysterious Tunguska Event—The Best of Carl Sagan's Cosmos.

Arthur C. Clarke's Mysterious World—Tunguska explosion.

The Tunguska Mystery 100 Years later: *Scientific American*.

INDEX

A

Academy of Sciences, USSR, 143, 164
Alvarez, Luis, 10
Anfinogenov, John, 56
Anfinogenov's butterfly, 56
Apollo Asteroids, 76
Apollo Moon missions, 40
Aristotle, 44-45, 107
Asimov, Isaac, 3-4, 163
Astapovich, Igor, 29

B

Ball lightning, 31, 141, 150-151
Barringer crater, 52
Beta Taurid meteors, 74, 150, 161
Bidyukov, Boris, 55
Blast/Conical Waves, 13, 17, 57
Bologna University, 87, 136, 146
Bostick, W.H., 151
Boyarkina, Alena, 152, 155, 160
Bronshten, Vitaly, 64, 69, 84

C

Chicxulub crater, 33, 135
Clarke, Arthur C, 3, 163-164
Collisional ejection of Moon, 43-44
Comet Encke, 4, 69, 73, 79-80, 87, 93-97, 100-101, 126, 155
Coronal mass ejections, 96, 103, 109, 113, 153

D

Dinosaurs extinction, 10, 33-35, 135
Dmitriev, Alexey, 151, 160

E

Einstein, Albert, 3
Eye witnesses to Tungus event, 58

F

Fast, Wilhelm, 3, 55
Fast's butterfly, 56
Fesenkov, Vasily, 69, 83-84, 138
Filimonova meteorite, 50
Forest, devastation, 22, 94, 121
Fowler, Prof, 153

G

Galileo, 3, 45, 108, 129
Geomagnetic storm, 152, 156-160
Geometeors, 145, 150

H

Halley, Edmond, 82, 98-100
Halley's Comet, 78, 82, 88, 99, 132, 149
Hiroshima, 49, 53, 65-66, 91
Hollis, H.P., 153

I

Ilyin, Anatoly, 57
Iridium, 134-135, 150
ITEG, 54, 56-57, 59, 65, 67, 135, 157, 160
Ivanov, Kim, 157

J

Jet Propulsion Laboratory, 8

K

Kasantsev, Alexander, 65
Kepler, Johannes, 45
Kirkwood Gaps, 131
KMET, 53, 56
Konenkin, Victor, 146
Krinov, Evgeny, 24
K/T boundary, 34, 135
Kuiper Belt Objects, 5, 130
Kulik, Leonid, 3, 13

L

Lake Cheko, 7, 72, 136, 147, 162

M

Magnetic anomaly, 136
Magnetosphere, 65, 110-111, 116-117, 151-152, 155, 158-160
Maunder, E.W., 153

N

NASA, 5, 8, 40, 69, 99, 110, 125, 127, 164
Newbegin, G.J., 152
New Horizons mission, 99
Newton, Isaac, 38

O

Ogdy, Tungus god, 36, 61, 137
Oort Cloud, 5, 82, 95, 97-98, 125, 141

P

Plasmoid, 11, 118, 151, 155, 160-161
Plekhanov, Gennady, 53
Pluto, 4-5, 43, 90, 98-99, 130, 163

R

Romeiko, Vitaly, 64, 87

Rubstov, Vladimir, 163

S

Sagan, Carl, 3-4, 164
Sidoras, Saulas, 160
Sikhote-Alin meteorite, 136
Suslov, Innokentiy, 15

T

Taurid Complex, 73, 75-76, 78-80, 91, 93, 100, 126, 161, 164
Tesla, Nicola, 140
Thermoluminescence, 55, 135
Tungussky Reserve, 47, 55
Tycho, 45, 128, 144

V

Vasilyev, Nikolay, 53
Vernadsky, Vladimir, 14, 69
Voyager missions, 4, 99, 110
Voznesensky, Arkady, 18

W

Weather at Tunguska, 17, 21, 25, 29, 44, 50, 57, 96, 141, 152, 154, 164
Weber effect, 81, 160
Whipple, Francis, 69
Whipple, Fred, 76, 81, 101

Z

Zenkin, Igor, 57
Zhuravlev, Victor, 53, 160
Zolotov, Alexey, 55

Printed in the United States
By Bookmasters